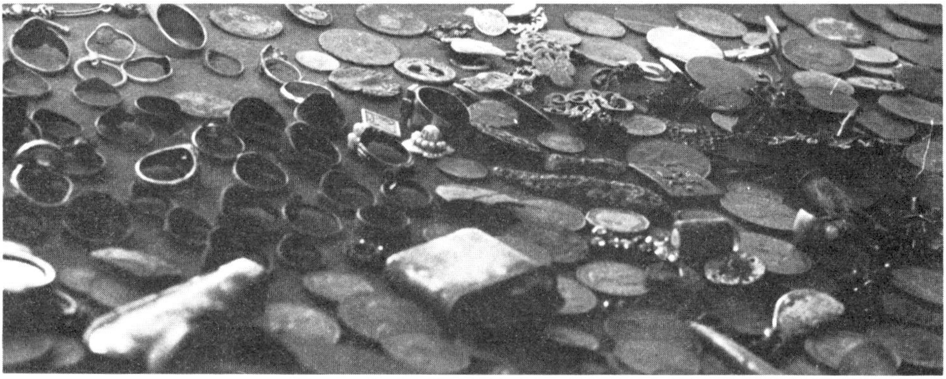

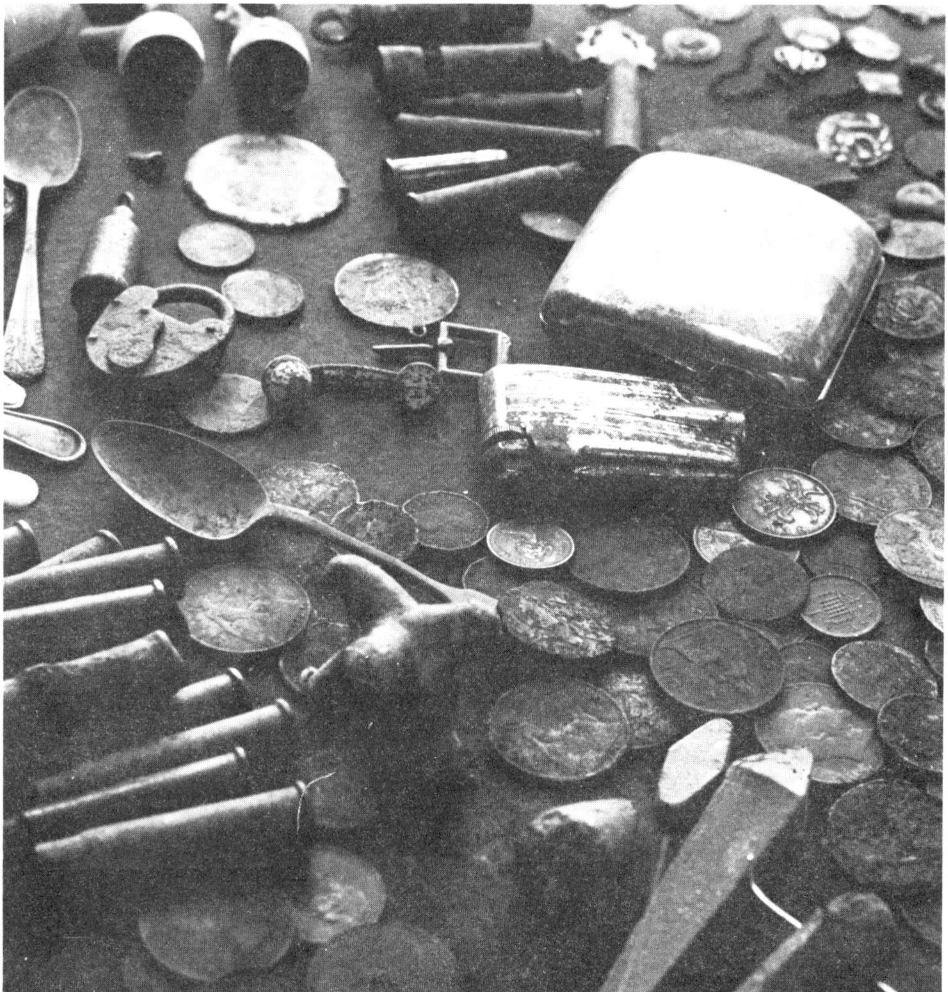

Overleaf: part of the large collection of coins, badges etc., found by Bob Renfrey professional treasure hunter.

TREASURE HUNTING

All rights reserved. No part of this book may be reproduced or transmitted in any form, without permission in writing from the publishers.

ISBN 0 905447 14 X (Paperback only)

ISBN 0 905447 13 1 (Hardback only)

© Copyright E. Fletcher and M. A. B. Publishing

Text set in 11 pt. Photon Times,
printed by photolithography, and bound in Great Britain
at The Pitman Press, Bath

TREASURE HUNTING

PROFITABLE FUN
FOR THE FAMILY

by
Edward Fletcher

M. A. B. Publishing
801 Burton Road, Midway,
Burton-upon-Trent, Staffs DE11 0DN
England

The Author

Author Edward Fletcher with his wife Doreen and some of the "treasures" they have found during a lifetimes interest in the hobby.

Edward Fletcher is known to treasure hunters and bottle collectors throughout the world. He has written many books on the subjects of treasure hunting, bottle collecting and rockhounding.

His success is due not only to his abilities as an author but also to the fact that he actually goes out and does the things he writes about. He is actively involved in treasure hunting and bottle collecting clubs and he has done much to promote these hobbies in Britain.

Contents

Introduction	7
A family hobby	11
Fun for the kids	17
Business with pleasure	23
Profit from discrimination	31
Beachcombing	40
Treasure hunting clubs	47
Treecombing	49
Striking it rich	56
Dump digging	63
Rockhounding	70
Searching for Gold and Pearls	73
The right approach	75
Thoughtful research pays	88
Useful information	93
Acknowledgements	96

Cover Photograph

Cover photograph (Gordon Litherland).

A selection of treasure hunters finds from Gordon Litherland's and Bob Renfrey's collections. Listed are a few of the items:

1. Cobalt blue codd bottle (dug).
2. Brass barrel tap (found by detector).
3. 17th century Bellarmine jug (found in river).
4. Onion bottle (c.1690) (from the sea).
5. Cobalt blue hamilton bottle (dug).
6. Victorian doll's head (dug).
7. Face ink (found in loft).
8. Clay pipes (dug).
9. Coins (found by detector).
10. Horseshoe (found by detector).
11. Vespa (match box) (dug).
12. Pot-lid (from chemists cellar).

Introduction

Major developments have taken place within the world of treasure hunting over the last few years. Perhaps the most important of these changes has been the astonishing growth in popularity – from an obscure "profession" with a handful of eccentric and rather secretive devotees, to a popular family hobby avidly followed by tens of thousands of weekend amateurs.

This universal adoption of amateur treasure hunting as a rewarding and enjoyable weekend pursuit by large numbers of people from all walks of life is directly responsible for the second great change that has taken place during the past few years: the development and marketing of a breathtaking bazaar of treasure hunting equipment, much of it sold through a network of "treasure hunters' shops" scattered throughout the land and providing almost everyone who decides to take up the hobby with a *local* dealer who can offer his customers a choice of equipment over a wide price range while at the same time giving personal service, friendly advice, and useful information on the local treasure hunting scene.

How different things were back in 1969 when I was first bitten by the metal detector bug. At that time there was in Britain only one detector manufacturer who shared the small "eccentric" market with a single importer of American machines. Both were based in London and potential customers had a simple choice – a visit to London or a mail-order purchase. Today newcomers can choose from more than fifty different metal detectors made by dozens of manufacturers at prices ranging from under £20 to more than £300. There is also available an increasing amount of accessory equipment including such fascinating gadgets as sea-searcher magnets, suction dredges, coin-extracting tools, gold pans, riffle boards, coin cleaners, and more. Every county in the land now has a choice of shops where beginners – perhaps attracted by reading a book on the subject, or hearing of a friend's successes at finding old coins, or spotting advertisements in such magazines as "Exchange & Mart", "Gemcraft" and "Treasure Hunting" – can see all the "hardware" on display and buy some of it after a face-to-face demonstration by the dealer.

Another striking difference between "then" and "now" – one which makes me very happy indeed to be able to claim some of the responsibility for bringing the hobby to the attention of a wider public – is the astonishing success rate that has already been achieved by the hobby's following, even though (despite the numbers of people already involved) treasure hunting is still in its infancy in Britain and virtually unknown in continental Europe. Few outside the hobby are aware of just how much has so far been found. I do not doubt for a moment that when I state that hundreds of thousands of coins, items of jewellery, military badges, medallions, antique weapons, and other "collectables" have been recovered, many outsiders will dismiss my claim as exaggerated. They will scoff when I write that less than an hour before I sat down to type this page I received a telephone call telling me of the successes of a 13-year-old Carlisle schoolboy who has just located – with a detector costing less than £40 – the beginnings of a scattered hoard in a field near his home. At the time of writing 123 coins have been recovered, with the bulk of the hoard yet to be pin-pointed, (sceptics can verify the facts by consulting Carlisle newspapers which, I am told, will be covering the story in detail). That is just one of hundreds of hoards found by amateur treasure hunters in the past few years – gold hoards, silver hoards, misers' hoards, robbers' hoards, and the personal savings of a tiny fraction of the people who, during the past 2,000 years, have put their money in the ground for safe keeping and been unable, due to death or upheaval, to recover it. Only now has 20th century science and "electronic magic" enabled us to begin to find this fabulous buried treasure which, if it is ever totally recovered, will make the treasures of the Inca civilization seem like the coppers in the tray of a Victorian match seller. . . .

It is not merely an increase in the numbers of amateur treasure hunters that has led to the recovery of so many single finds and large hoards. I doubt if so much would have yet been recovered if there had not been in recent years a frenzy of activity by the "back room boys" whose achievements in improving the performance of metal detectors deserve far more credit than has yet been given. It is a fairly simple task for anyone with a knowledge of electronics to design a metal detector that will find an old penny six inches beneath the surface of the ground. But it is ten times more difficult to improve that depth penetration to seven inches, and ten times more difficult again to add even half an inch to that seven. Hats off then, to those dedicated researchers and inventors who have, in only three or four years, achieved depth penetrations for single coins in excess of ten inches.

Yet more laurels must be heaped on their heads for those enormous strides in detector technology which have given us water immersibility, drift-free coils, discrimination between ferrous and non-ferrous metal, wide-scan search heads, and pin-point accuracy when locating small coins. If they can maintain this galloping rate of invention for just a few years more I am confident that we will see the production of detectors capable of finding only gold and silver coins at depths in excess of two feet. (In fact experimental models have been produced).

That then is the present state of the hobby in Britain. A success story so far as those who were in at the start are concerned; those whose treasure hunting activities have evolved as the equipment, the expertise, and the popularity have grown. But what of the beginners? – the newcomers buying their first metal detectors who are unfamiliar with the "jargon" of treasure hunting; who know little or nothing about induction balance, pulse, ground exclusion, very low frequency, or the many other technical terms bandied about by those of us who have grown up with treasure hunting in our veins? Regrettably they are rather poorly served so far as written information on treasure hunting equipment and techniques is concerned. In spite of all the popular interest mentioned above, beginners in Britain have only a choice of three or four books about the hobby.

It is the answering of that basic beginner's question "can I make a profit from Treasure Hunting?" that is the main purpose of this book. In the chapters that follow I have attempted to answer the question from every angle and in a way that will provide a satisfactory answer for every beginner who reads the book. I stress "from every angle" because there are many ways of looking at the question. An answer that satisfies one reader may not satisfy another. A schoolboy or girl would, no doubt, find it profitable to hunt 1p and 2p pieces lost only a few hours beforehand on a public beach. But such "pocket-money" activities would certainly not satisfy an experienced coin collector with a passionate interest in hammered silver coins. Such a reader requires a different answer and different treasure hunting equipment. I have, therefore, expanded my advice to include details of the various types of equipment needed to achieve different kinds of profitability. If readers follow this advice by visiting or writing to the various dealers and manufacturers they should soon be on the happy road to PROFITABILITY.

A Family Hobby

As a family man it is my belief that the key to family happiness is TOGETHERNESS. If husband, wife, and children have a shared activity and interest then family harmony follows automatically. What better way then of achieving togetherness than by taking up amateur treasure hunting as the family hobby? Here is PROFIT beyond price – the mutual enjoyment of a pastime that gets the entire family out-of-doors, following a fascinating trail of buried history. Put the idea to your kids and if they don't give out "whoops" of joy I'll eat a copy of this book. It may take a little longer to convince your wife that this is the best idea you have had in years, but when you tell her she need no longer be a golf or fishing widow; that you propose to devote all your free weekends to family treasure hunting; and that there is every possibility that within a few weeks you (and she) will be unearthing lost rings, bracelets, necklaces, ear-rings, and brooches, she, too, will warm very readily to your bright idea.

On a more serious note, there is absolutely no doubt that children who are part of a family where a father or mother takes an interest in relics of the past very quickly begin to take an interest themselves in history, research, museums, coins, antiques, and many other off-shoots of treasure hunting. Time and again I have been told by parents who have involved their children in family treasure hunting that they have seen a marked improvement in standards achieved at school in History examinations – also, incidentally, in Geography (via map reading), in Woodwork and Handicrafts (via the making of display cabinets for finds), in Science (via metal detector circuitry), and in English (via the reading of Victorian authors including Dickens and others for background information about 19th century society).

It need not cost a fortune to equip yourself for family treasure hunting. Certainly, you do not at the outset require a metal detector for each member of the family. A far better start can be made by purchasing one reliable unit that also has an adjustable handle which can be reduced in length until the detector is of a convenient height for use by the youngest member. Obviously you should also look for a model with simple controls which a child of ten or twelve can learn to operate under your supervision. (There are at present dozens of models on the market that meet these requirements.)

You will save much frustration and time-wasting on your first family outing if you master the basic techniques of tuning and using your new detector before group activities commence. Take it out alone on several evenings before the "big day". Go to a local common or public footpath and practice tuning the controls. Make sure you are quite capable of finding coins by the end of these practice sessions and you are now ready to tackle the first family hunt.

Children old enough to handle sharp tools safely should each be given a broad-bladed knife or sharpened trowel, and it should be mutually agreed that for the first hour Dad or

Sites of Victorian pastimes can be interesting for the treasure hunter.

Mum will do the "locating" while the kids take it in turn to unearth the finds. (If the site chosen for the first hunt is a grassed common or field make sure the children know how to cut a neat cap of turf by letting them practice in your back-garden.)

I recommend a well-used common, recreation ground, or unlandscaped park for your first outing. A beach is not so good a choice because coins sink rapidly in sand; if you are still fairly inexperienced at coin hunting you are likely to make far more finds on well-used grassland on your first outing – and it is most important to keep the rest of the family involved with extracting lots of buried objects on this first trip.

The key to initial success on the first half-dozen family hunts is to ensure that every member of the family participates fully in all the activities. Don't "hog" the detector simply because you prefer locating finds rather than extracting them. Give Mum *and* the kids a chance to practice the techniques of handling the machine while you unearth their finds. Be lavish in your praises of their successes – even if all your wife finds on the first day is a battered George VI halfpenny. If very young children are involved make a "big thing" even of their bottletop finds. Explain that although the bottletops are worthless they have been buried for ten, twenty, even thirty years. Give toddlers the job of carrying these junk items until you come upon a litter basket where they can be ceremoniously dumped with a comment from you that even though the family hasn't yet found a fortune it is doing a great job of keeping Britain tidy. If you return home at the end of your first "togetherness day" with only half-a-dozen coins but with every member of the family in agreement that it was an enjoyable outing you are well on your way to success.

After two or three forays as described above you can begin to consider more serious expeditions in search of Victorian (and earlier) finds. Now is the time to involve all members of the family in RESEARCH, the "open sesame" to valuable coins and relics. You could very profitably

Typical Victorian scene of river bathing. Location of such riverside sites can produce coins, buttons and small personal belongings of the period.

spend a wet Saturday morning down at the local library browsing through old books on local history or looking through files of old newspapers. EXPLAIN to your children why this research is so vital; that you must establish where Victorians spent their leisure hours if you are to find the items they lost on picnics and outings or while playing games. Older children will know at once the sort of information you are seeking; but it is worthwhile spending time to explain to youngsters how scanning the "Lost and Found" columns of 19th century newspapers can provide vital clues to where the townsfolk spent their weekends in those days. If while scanning old newspapers covering a period of six months you come upon three or four adverts which state that jewellery, purses, etc. were lost on "Blogg's Common" then obviously "Blogg's Common" is where the family should hold its next treasure hunt.

An aspect of research which all youngsters thoroughly enjoy is the questioning of old people about where they played as children. If you have Grandmas and Grandpas in your family the children should encourage them to talk about "the good old days". The very useful information they might gain includes where they played as children, where the traditional toboggan runs were, where Easter eggs were rolled, where fairs were held, and which local woods and commons were the most popular for weekend outings. Older children in the family might compile a questionnaire and carry out a neighbourhood survey, putting these questions to every old person in the district. If the old people interviewed are over eighty and they can recall their childhoods vividly you will quickly build up a most useful file of information on local sites – information that is certain to lead to exciting finds of Victorian coins and relics.

As you become more experienced at using your detector and finding productive sites you will soon begin to build up an impressive collection of finds. Don't simply lock them away in drawers and cupboards. Encourage the kids to display them – on walls, shelves, fireplace surrounds – anywhere that keeps them in view. They will then serve to tempt the family out on the "treasure trail" once more. Your modern coins should be saved in a family moneybox until you have sufficient to buy a second detector. CHOOSE IT CAREFULLY. You will considerably increase your success rate if, instead of buying a cheaper detector than the one you already own, you pass on the first unit to another member of the family and buy a more sophisticated model to replace it. Although this will mean waiting a little longer until you have found the cash to buy it, IT PAYS IN THE LONG RUN. With two reliable detectors you will at least double your productivity and speed the day when every member of the family has his or her own unit. What you do with your modern cash finds after that I leave to your imagination – BUT don't bury them in the ground. There are other folks at work with detectors. ...

Once your reputation as the local treasure hunting family gets around the neighbourhood, or as a result of your children taking some of their finds to school, you will begin to receive enquiries from interested friends who are also thinking of taking up the hobby after seeing your successes. When the ball starts rolling in this way keep up the momentum by forming a local treasure hunters' club. If all members pool their resources when carrying out research and also share the cost of travel to more distant sites you will find that your collection actually increases rather than decreases as more people in the neighbourhood join in. Thus, the more involved your family becomes with amateur treasure hunting the greater will be your profit from the hobby.

Treasure hunting on the beach can provide profitable fun for the family on holiday.

Fun for the Kids

THIS SECTION IS FOR READERS UNDER 14 YEARS OLD ONLY. ADULTS PLEASE SKIP TO NEXT SECTION.

What is the difference between a piece of elastic and pocket money? ANSWER – one stretches, one doesn't – and you don't need me to tell you which one doesn't. No matter how hard you try it is just about impossible, thanks to inflation and the rising cost of sweets, comics, and everything else, to have even a few coppers left out of your weekly allowance by Monday morning. So how on earth are you ever going to get together enough money to buy a metal detector which will cost you up to a whole year's pocket money?

Well, let me tell you about a boy I know who managed the trick – even though his weekly pocket money amounted – before he owned a metal detector – to only 50p per week. He's quite an ordinary boy; the type who usually gets described by teachers as "fairly good" – fairly good at Maths, fairly good at English, fairly good at Sport, etc. About the only thing he's better than fairly good at is WANTING. He's brilliant at wanting, and when he wants something badly enough NOTHING is allowed to stand in his way.

About a year ago this boy (I promised I'd keep his name out of the story, hence my repetition of "this boy") saw an item in a television programme about treasure hunting. Before the end of the programme he had been bitten by the WANTING BUG. He wanted a detector so badly it hurt him to think about it. Actually, the real reason it hurt to think about it was that he had only recently "wanted" a bicycle and his Dad, who gave in on that occasion, had told him quite flatly that he'd get no more presents until Christmas. As it was then only the middle of May you can imagine why wanting a metal detector caused him pain....

Well, as I've said, this boy is brilliant at wanting. As soon as he learned through reading advertisements in various magazines that he could buy a detector for about £20 he set about the job of saving £20 with all the determination of a mountaineer about to tackle Mount Everest. First he did the obvious things such as odd jobs for Mum and the

This young digger is showing off his "prize-of-the-day" a ginger beer bottle.

A boy is fascinated by a display of old bottles at the Burton-on-Trent bottle show.

neighbours. These tasks earned him about 75p before he ran out of jobs that needed doing. Then he hit on the idea of asking his Dad if he could search the loft in the hope of finding a valuable oil painting or something else that might be turned into cash. His Dad thought the idea was crazy, but as he was about to put some insulating material in the loft he agreed to "have a look round while up there". Alas, all he found were some old comics and a couple of Dinky toys with their rubber tyres missing.

Not wishing to look a fool, the youngster put a brave face on and said he'd take them down to the local flea-market on the next Saturday in the hope that he'd get a few pence for them. Secretly he didn't expect to raise a penny on that old junk – BUT – he was in for a pleasant surprise. It turned out (he discovered as he strolled around the stalls at the flea-market) that there were people actually prepared to pay hard cash for such items as old comics and Dinky toys without wheels. Even his Dad was surprised when the lad returned home with a pound note....

His Dad took another look around the loft but it was bare of more valuable "junk" – or if there was any it was now hidden beneath the insulating material. Undetered, my young friend next persuaded various aunts, uncles, and grandparents to hunt through their lofts. He even talked them into sorting the contents of cupboards and garden sheds – a move that paid off handsomely. His collection of "junk" at the end of a week-long search amounted to two old flat-irons, a World War II gas mask, a badly torn oil painting in a heavy ornate frame, a pile of old 78 rpm records, a dust-covered bellows camera that didn't work, some faded postcards in a battered album, and an old biscuit tin (empty) with a colourful picture on its lid. He did also find an old photograph album containing pictures of his Dad in his schooldays. When his Grandma saw it she was delighted. She had been trying to find it for months and certainly didn't want to part with it – but she did give him 50p reward for tracking it down.

Well, the lad had little difficulty in disposing of his finds down at the flea-market. Even the torn painting was bought for its ornate frame and when he'd finished his selling spree he had £12 – plus the 50p his Grandma had given him. ... More than half way to a metal detector.

For the next few weeks the only way he could add to his savings was by cutting out all luxuries and holding on to a copper or two from his pocketmoney. Then, quite by chance, he learned that a group of boys at his school had begun to collect old bottles and were digging them up on an old refuse dump at the rear of a nearby recreation ground. On the following Saturday morning, having been told by the bottle collectors what he needed, he borrowed a fork and a shovel from the garden shed and made his way to the old dump. To his surprise he found there were twenty or more people already at work – some of them grown-ups who had dug holes several feet deep. He found a hole that had been abandoned by an earlier digger and set to work raking the walls. He was rewarded with quite a few bottles of various shapes and sizes, but was told by an older boy who joined him in the hole that they were worthless; he must look instead for bottles with marbles in their necks and stone bottles with trade marks stamped on their surfaces. A full and rather dirty day produced only one each of these for the lad. But he did also find a round pot lid with black writing beneath the glaze and was delighted when an adult digger offered him £5 for it.

During the next few weekends he managed to dig up several more marble-stoppered bottles and one or two of the stone types which, by then, he knew were old ginger beer bottles. Although he could only manage to get 25p each for these at the local flea-market he eventually found sufficient to push his savings over the £20 mark. At last he had the price of his first detector. ...

Now I'm not suggesting that you do exactly as this boy did. You may not have lofts full of valuable junk or old rubbish dumps on which you can dig up saleable bottles. What I am saying is that you don't have to wait patiently for birthdays or Christmas to own a detector. Think about it

Children using a metal detector by the side of an old gravel pit.

and I'm sure you will come up with dozens of ways in which you could raise the cash – hold a jumble sale, sell your stamp collection (with your parents' approval of course), keep up an "odd job" service until you reach the target. It's easy – if you really set your mind to it. . . .

Let me tell you a little about how my young friend now uses the detector he worked so hard to buy. It's what is known as a beat-frequency model (BFO is the term used by experienced treasure hunters) and it does NOT penetrate the ground very deeply when locating coins – as the lad found out when he tried hunting Roman coins on a ploughed field. He failed to find anything so he turned his attention to a local beach where, to his delight, he was able to pick up lots of modern coins – mainly 2p and 10p specimens that had been lost very recently. He also visits parks, commons, and other places where money is lost by the large number of people who use these sites for picnics, games, etc. My young friend regularly finds up to 50p per week (over £1 on holiday weekends) with his inexpensive detector. These finds have enabled him to start saving for a better detector without having to miss out on sweets, cinema visits, and other luxuries for which he uses his regular pocket money. Next year he hopes to "trade-in" his first model and add to the price he gets for it sufficient "found" money to buy a metal detector costing almost £50. YOU COULD DO THE SAME IF YOU HAVE THE PATIENCE TO STICK AT THE TASK UNTIL YOU SUCCEED. THERE IS PROFIT FOR KIDS IN TREASURE HUNTING – EVEN WHEN THEY USE DETECTORS COSTING LESS THAN £20.

Once you own a metal detector you can of course start building up all sorts of interesting collections with your finds. There are thousands of badges lost in the ground – military *and* non-military specimens which, when cleaned up, make very interesting displays. There are also numerous metal buttons, pen-knives, pins, fishing weights, buckles, and items of jewellery. Toys can also be found – all sorts of toy cars and lead soldiers have been lost on commons, playing fields, and public footpaths. A metal detector will help you find them – even when they are covered by several inches of soil.

Business with Pleasure

Every year police forces throughout Britain receive tens of thousands of reports concerning items of jewellery lost in public places – beaches, parks, sports grounds, public footpaths, and elsewhere. Many come from women who have lost wedding and engagement rings of values ranging from fifty to five hundred pounds, which have slipped from their fingers as they swam, or played tennis, or stumbled while walking, or slipped when crossing a stile. A smaller percentage come from men who have lost watches, lighters, wallets, tiepins, and cufflinks under similar circumstances. If one accepts an average figure of 20,000 reported losses each year and sets an average value of £50 on each article their total value can be estimated at ONE MILLION POUNDS EVERY YEAR. Small wonder that insurance premiums are so high – but the true figure for losses of personal items such as this is much greater. Less than 25% of those who lose rings on public beaches actually report the loss to the police simply because they rate the chances of recovery as so slim it is just not worth the effort of making the journey to the police station. As for the untold thousands of people who lose jewellery and other valuables in their own gardens while weeding or mowing their lawns, they either feel too foolish to report the loss to the police, or they take the optimistic view that the lost items will "turn up one of these days". Add these unreported losses to the figure already mentioned and – incredible though it may seem – we can conservatively estimate that gold and silver with a retail value of TWO MILLION POUNDS is lost every year. There is, of course, a percentage of recovery – but it is small; certainly less than half.

Readers who find these figures difficult to credit might look at the question of jewellery losses in another way by asking themselves if they or members of their families or their immediate neighbours have lost pieces of jewellery during the past twelve months. I am sure the answer will be YES – and that applies to EVERY reader. I expect this book to sell at least 20,000 copies during the next year, a figure which, when multiplied by an average ring value of £50, brings us to an estimated loss of £1,000,000 per year. . . .

As a detector owner you can profit greatly from all that lost gold and silver either by recovering some of the lost items for their owners or by claiming recovered pieces when owners cannot be traced. If you live in a popular seaside resort and you own a detector with a good depth penetration suitable for beach work you can very quickly build up a profitable part-time business as follows: Have a few business cards printed which introduce you as a finder of lost rings. (e.g.: "John Smith. Search and Recovery Services. Specialist in locating jewellery lost on local beaches. Moderate fees. Phone Newtown 11234. Anytime.") Call in at your local police station and leave a card, asking the duty officer to refer to you anyone who calls to report a lost ring. Place some of the cards in newsagents' windows around the seafront area of the town; also hand out as many as possible to taxi-drivers, hotel receptionists, and the girls behind counters at the local tourist information office.

Next call at your local newspaper offices and place a small ad. with similar wording to your business card in the "Lost and Found" column. Run it on several days each month throughout the summer season. If you are the first treasure hunter in your area to hit on the idea of such a service it is very likely that one of the newspaper's reporters will spot your advertisement and consider it an interesting item for a feature article. This will provide you with valuable free publicity so do all you can to make the story interesting and informative. Tell the reporter you are willing to attempt searches for anything and that your equipment can recover other things besides lost jewellery.

All you need to do now is sit by your phone and wait for calls to come in. . . . And come they will; thick and fast on those hot summer days when the beach is crowded with sunbathers and swimmers. You may even find yourself looking for two rings on the same stretch of sand for two different people on your busiest days.

A FEW WORDS OF ADVICE ON SEARCH METHODS, FEES, ETC: You will soon learn from frustrating experience that people who lose jewellery and other personal possessions are often unable to give you much useful information on where the item was lost. You will come to recognise a phrase such as "I lost it near the whelk stall" or "It must have dropped from my finger as I was putting the deck-chair in position", NOT as the place

Opposite: search service on older properties will probably produce finds from by-gone ages as well as the modern item you are searching for.

where the ring was lost, but as the place where the loss was first noticed. Before committing yourself to a search ask a few questions. You must try to narrow down the area of the search – first by establishing just how far and wide the loser has wandered on the beach during the day, then by establishing where the lost item was last seen. It is not easy to do either of these things because the loser invariably insists that he or she has "not moved an inch all day" or "quite definitely saw it only minutes before it was lost". I therefore strongly recommend that you hire yourself and your detector for an hourly fee rather than a fixed price for finding the object. Fees seem to range, among those treasure hunters who are already at work providing this service, from £3–£8 per hour, depending on the value of the lost item.

The only times when items lost on beaches or in other public places should be hunted for a fixed fee are when the owners actually saw the objects dropped from fingers, necks, etc. and you can thus limit the search area to a few square yards. The recommended fee is £10 per £100 in value.

Another most important point is to have an accurate description of the item before the search commences. This avoids embarrassment if you happen to locate a piece of jewellery that is not the subject of the search you have undertaken. THIS HAPPENS QUITE OFTEN ON BEACHES.

It is most important that searches for objects lost on beaches are carried out as soon as possible after the loss. If the tide comes in and covers the sand even once before you begin your search the item will become deeply buried. If the owner has delayed contacting you for more than twelve hours since the loss don't accept the job on a by-the-hour basis; you might spend ten hours searching and still fail to find it. Far better under such circumstances is to take the name, address, and telephone number of the owner and explain that you will get in touch if the object turns up during one of your searches during the next few months.

Searches in private gardens can of course be undertaken

Search services for insurance companies can introduce you to the owners of large estates.

days, weeks, even months after the loss with a very good chance of success because the lost item is likely to remain within the penetration range of your detector for a considerable time. But do satisfy yourself that the item was indeed lost in the garden before taking on the job on a no-find-no-fee basis. If you cannot satisfy yourself on this point stick to a by-the-hour scale of charges.

If you are prepared to go to the expense of buying the very best equipment – sophisticated IB units and (for some sites) pulse machines – you could build up an interesting and profitable nationwide service by undertaking search and recovery work for insurance companies. These companies pay out scores of thousands of pounds every year to owners of lost jewellery and they are very anxious to cut down the amounts paid in loss claims. Have some attractive letterheads printed and write to several of these companies, explaining the nature of your business and giving approximate fees for searches. In order to get established it may be necessary for you to undertake your first few jobs on a no-recovery-no-fee arrangement so *do* explain in your initial letter that you must limit the service to hunting for objects lost in private gardens or to hunting for items lost in public places where the owner can state with certainty that the item was dropped within a small area. It would also be wise to limit your service to an area within fifty miles of your home – at least until you have proved to the insurance companies that you can save them money. Once you have notched up a few successes you can ask for work on a by-the-hour basis and with a fee to cover your travelling expenses.

There is an added bonus in taking on jobs for insurance companies: Many of their clients are rich and live in big houses with large private grounds. Once you have introduced yourself to the client and he or she has seen your detector in operation it is often possible to obtain permission for a full search of the grounds in the hope of recovering coins and other relics. If you succeed in locating the lost jewellery you will probably be invited to carry out further searches.

Another profitable search service that can be undertaken even by an amateur treasure hunter equipped with an inexpensive BFO unit is the location of manhole covers for local councils. It is not widely appreciated that very few councils have town maps which give the exact locations of manhole covers. I was told by one Borough Engineer with whom I discussed this problem that his department knew "only to within ten feet" where the hundreds of manhole covers in the borough were located. This inaccuracy on his maps resulted in many of the covers becoming "lost", either during months when the ground was covered by snow or when civil engineering contractors accidentally buried them under layers of tarmacadam and asphalt. Without a metal detector the only way to locate a lost cover is either to shovel away up to one hundred square feet of snow or to dig random holes in roads and footpaths in the hope of striking the cover. Of course, quite a few councils have already purchased detectors which they use for this very job; but there are many which have no electronic aids for this sort of work – and even those that do would probably prefer to sub-contract the work to a specialist. Again, letters explaining the service you could offer should be sent out – to Town Clerks and Borough Engineers. You should also send letters to local civil engineering and public works contractors who often have similar problems. Building and scaffolding contractors might also be interested in hiring you and your detector to locate lost tools, scaffolding tubes, and other hardware. As with searches in private gardens, these jobs can result in permission being given to carry out searches for coins and other interesting items – once the people you work for have had the opportunity to appreciate your expertise.

I know of several experienced treasure hunters who have obtained permission to search farm fields where Roman and other coins have turned up by offering a free-of-charge service to farmers and market gardeners worried about damage to expensive machinery caused by large hunks of iron lurking a few inches beneath the soil's surface. An offer to clear a piece of land of such hazards is excellent "bait" when attempting to obtain permission for treasure hunts. Although I know of no treasure hunter who has yet tried it, I'm sure a similar offer made to the Parks Superintendent in your town would enable you to search in parks, recreation grounds, even on football pitches where costly damage is done to lawn mowers by stray pieces of metal. It would of course be essential to carry out a very careful search using lines and pins in order to ensure that every metal object (including coins) was removed. The most difficult part of such an operation would be to convince the Parks Department that you could do the job without damaging the turf. Why not ask the Superintendent to accompany you on a hunt so that he can see for himself how pieces of iron can be extracted without chopping large holes in the grass?

There is another way in which you can obtain permission to search the grounds of private houses, estates, even stately homes. If you were to write to all the owners of large houses in your district offering to recover antique coins and other relics on a "share basis" I'm sure this would lead to some very interesting searches. It is people whose families have lived for generations in the same properties who have the most detailed family histories – often including tales of eccentric ancestors who buried their fortunes without telling anyone where to find them. If even one of these tales you hear during your search services turns out to be accurate there could be BIG PROFITS in it for you. ...

I must stress once again the need for sophisticated equipment and very careful search methods if you are to succeed with a business offering recovery services. SUCCESS CERTAINLY BREEDS SUCCESS IN THIS GAME; if you become known as a successful searcher you will be swamped with work. But you will NEVER succeed if your idea of a thorough search amounts to little more than a trot around the grounds waving a cheap and nasty gadget three feet above the surface. If your aim is to succeed then get out and talk to as many experienced treasure hunters as you can possibly meet. Find out by asking them which detectors they use and what they recommend as the type of equipment required for search and recovery operations. When you've listened to a dozen or a score of opinions you can begin to make sound judgements – though you'll still be learning a year from now.

Profit from Discrimination

Victorian rings and jewellery.

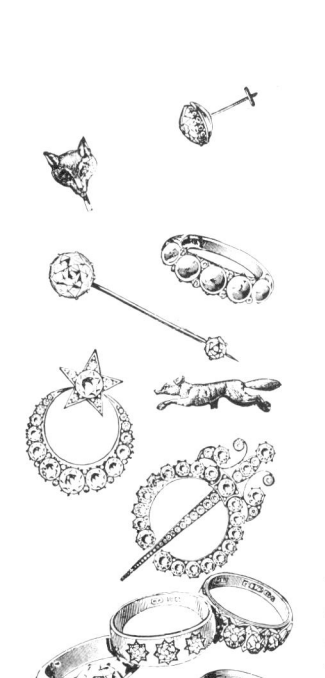

Every treasure hunter must consider where he will use his detector and what he is seeking and decide, on that basis, whether to buy a discriminating or non-discriminating unit. Each has its advantages.

Detectors capable of discriminating against silver paper, bottle tops, and *small* pieces of iron are among the most recent developments in detector technology. Their introduction to the amateur treasure hunting scene has been heralded by fanfares of publicity which have included such statements as "the answer to your treasure hunting dreams" and "positive junk elimination possible at last". Indeed, it would seem on first reading of some of the advertising blurbs that one need only wander around a piece of ground with a discriminator in one's hands for the coins and jewellery to pop up out of the soil and dance at one's feet.

The idea of discrimination is not new. The very first metal detectors – *mine* detectors as they were known – "discriminated" against all but the largest pieces of iron. I mean by this they could only find large hunks of iron. They did not "see" coins or jewellery – or bottle tops or silver paper. Later, as more people became aware of the vast numbers of coins and relics lying a few inches beneath the soil, detector circuitry and design were improved to the point where sensitivity was such that even coins the size of farthings could be "seen" by the electronics.

For the past half-dozen years competition between detector manufacturers has been concentrated almost exclusively on improving the depth at which detectors can locate the smallest coins. By 1975 the best-of-the-bunch had managed to push depth penetration to ten or twelve inches for an old penny. (I am writing here of genuine in-the-ground depths – NOT experimental measurements in test labs.) There were one or two models on the market which could beat twelve inches for an old penny – but they operated on pulse induction principles which gave extremely poor pinpointing capabilities and a sensitivity to small pieces of iron, although they are ideally suited for beach use, where their deep penetration can be utilized.

Twelve inches for an old penny may seem to the layman who knows nothing of amateur treasure hunting a rather shallow depth. But anyone who has been involved with the hobby for a year or two knows that a working depth of twelve inches will produce superb finds, and that for general coinshooting little more need be asked of a metal detector – except – that in addition to finding an old penny at twelve inches, the detector will ignore bottle tops, silver paper, hunks of iron, ring-pulls, and other "junk".

The problem for any manufacturer attempting this near-impossible feat of electronic magic is that in order to make a metal detector that does not "see" the junk he must at the same time reduce its ability to "see" the coins and valuables. Let's say a manufacturer who already makes a reliable IB unit capable of finding coins at twelve inches decides to introduce a new model with built-in discrimination. Being a reliable IB unit his detector already has excellent pinpointing capabilities and is not very sensitive to small pieces of iron (i.e.: less than the size of a farthing). But in the hands of all but the most highly skilled operator who has used the same model for a long time and whose ears have become attuned to its "whispers" and foibles, the detector gives the same signal for penny-sized pieces of silver paper as it does for coins. To overcome this the manufacturer reduces the sensitivity until the detector cannot "see" these pieces of silver paper. The detector operates just as before – except that its maximum depth penetration for an old penny has now been reduced – from twelve to six inches or less.

It is on this point – that discrimination also reduces ability to find some coins – that beginners (and quite a few more experienced users) are confused. This was vividly illustrated to me a little time ago when I took an afternoon off and went to a site about twenty miles from my home which produces excellent coins from all periods from 1750 to modern. It has been well "picked over"; indeed, it was the site of a recent meeting of the Weekend Treasure Hunters Club when several hundred coins were located there. I arrived at the site to find another enthusiast already at work. He was using a sophisticated instrument that had both discriminating and non-discriminating circuits and his method of working was as follows: He set out lines-and-pins covering about ten square yards and first went over the area in non-discriminating mode, marking with the heel of his boot those spots where he obtained a signal. Having covered the search area in this way he then switched to "discriminate" and went over the locations once more. If he obtained no signal he passed on without extracting the object. He should have followed the golden rule when using this type of detector to DIG when you have re-checked your first finding and not received a signal. The method worked very well; he already had a pocketful of coins (Victorian–modern) when I arrived and was very busy on a newly marked-out area. We chatted for some minutes as he explained his method and continued to pull out coins. He then noticed that I was using a non-discriminating IB unit which he was anxious to see in operation. On his invitation I re-located the objects he had declined to dig up and chose one on the extreme limits of my machine's sensitivity. (I must admit here that I liked the "sound" of it – a non-iron "whisper".) He checked the spot again, got a signal in non-discrim. mode, but again eliminated it when he flicked the circuit to "discrim." Pulling out my sheath knife I cut a cap of grass around the signal and probed the soil beneath. At the full depth of the blade I felt the touch of metal and burrowed with my fingers. To my delight out came a George III half-penny dated 1773.

In my opinion he also *missed* the most important lesson to be learned: that with a single find using a non-discriminating detector I had found a coin older by almost 100 years than anything he had found in an entire morning's search.

I am quite sure there is a very bright future ahead for those manufacturers who persevere with this type of circuitry. I shall be first in the queue on the day a manufacturer brings out a discriminator that can do its job at twelve inches – and give me pin-pointing accuracy, stability, ease-of-handling, and all-round usefulness as a treasure hunter's tool. I'm sure that day is not far off; it may even be upon us before this book has been on sale for a full year. But I am not a soothsayer. What I have written above refers to discriminators on the market in April, 1977 — as anyone who tries to throw my words back at me two years from now will be told.

Despite some disadvantages present discriminating detectors are most useful instruments. There are thousands of sites where a discriminator can be used with great success and handsome profit. Although amateur treasure hunting is primarily concerned with finding Victorian and earlier losses, it would be foolish to disregard the potential for finding modern, spendable cash and modern, valuable jewellery simply because these are 20th century losses. The truth of the matter is that very many of the deeply buried old coins found by amateur treasure hunters have little numismatic value; it is the thrill of the find that spurs us on to finding them, not their values as shown in a coin catalogue. If it's money you are after then present-generation discriminators have much to offer. Take one to a local common or a local wood which is very well-used for picnics and outings and you will certainly find plenty of spendable cash and substantial amounts of modern jewellery. If you tried using a non-discriminating detector on sites like this you would locate one hundred pieces of near-surface silver paper for every coin recovered; but with discrimination at no more than four inches you can eliminate 90% of this junk and find 50–100 modern coins during an eight-hour search. With a little luck two or three of these will be 50p pieces, half-a-dozen will be 10p pieces, and the remainder will be coppers. That would give you £3–£5 per day — all profit if you accept the pleasure of the search as payment for your efforts and you do not have to travel long distances to get to the sites. An old-aged pensioner who can find half-a-dozen sites like this within a few miles of his home (not an impossible task in large cities), could work them forever and make £18 – £30 per week — with the possible bonus of rewards for the recovery of lost jewellery. A child could do the same, with a little less success due to lack of concentration. An adult — perhaps trying to get together the price of a better detector — could

Medieval horseshoes – these are usually found in the country on old farms.

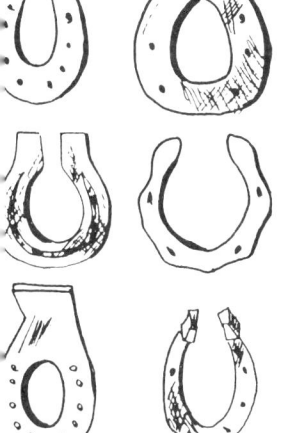

at least equal the performance of the old-aged pensioner.

Our American cousins have a name for these highly productive picnic and tourist spots. They call them MONEY PARKS – a term I consider worthy of adoption by the British amateur treasure hunting fraternity because our "money parks" hold even more modern cash than those on the other side of the Atlantic, due to the overcrowding of our little "treasure island". I have yet to find a means – short of taking them out for a day – of convincing non-treasure hunters that when I write that there are millions of post-1900 coins in the ground on sites such as this throughout Britain I am not exaggerating. Readers who already know that my statement is true must have had similar difficulties when talking to friends and neighbours about the hobby. I do not propose to use a long string of adjectives here in an attempt to persuade "non-believers" that I speak the truth. If they doubt my word let them find a local treasure hunter who is already using a discriminator on "money parks" and accompany him on a day's hunt. Seeing is believing, they say. Well, the results of "seeing" beneath the ground on one of these sites have to be seen to be believed. . . .

There are other sites on which the depth of discrimination we have at present can be very profitably employed, the best being tidal riversides where one often finds large amounts of iron nails and similar junk mixed up with antique coins and relics. I am going to presume that you have already mastered the techniques of "eyes only" searching and that you fully understand the theories behind the movement of coins by tidal currents.

Suppose you have found a spot on the foreshore where interesting coins and jewellery have been discovered beneath nail clusters. First remove all the visible finds and as many as possible of the nails and other pieces of junk. You now need a shallow, plastic bowl – as sold in many supermarkets. It should be approximately four inches deep and as wide as possible. Search the foreshore until you find a large flat stone or a log of wood and carry it back to the

After picking up all the "eyes only" finds, dig the foreshore material and place it in your plastic bowl.

Pass your discriminator over the material. A signal indicates non-ferrous metal in the bowl.

spot you are working. Place the stone or log on the ground and stand the plastic bowl on top of it. Using a small fork or shovel you must now scoop up the foreshore material – in the area where you have already made "eyes only" finds and fill the plastic bowl. Next switch on your detector (discriminating mode) and pass the head across the surface of the material in the bowl. If you get no signal empty the bowl and re-fill it with more material from the foreshore. Try another pass with the detector. If you now receive a signal you can say with certainty that the bowl contains a non-ferrous object – possibly a coin. Remove half the material and check the contents of the bowl once more. If you still get a signal the object remains in the bowl and should be easy to pin-point as you stir the material with your hand.

By using this plastic bowl method you overcome the problem of discriminators – inability to locate objects at depths much beyond six inches. You can work the same area of foreshore down to a foot or more, confident you are unlikely to miss old coins at great depths.

It can also be profitable on many tidal riversides to cover large areas of the foreshore with a discriminator to pick up the numerous pieces of non-ferrous scrap metal one finds there. This is an especially useful exercise when you plan a more detailed search for coins and relics later; by picking over the area initially to remove these objects you reduce the number of non-productive signals received when seeking coins. ALL the non-ferrous scrap should be retained and sold to a scrap merchant when you have sufficient weight. Copper, brass, and aluminium are the commonest metals found during these searches; some discriminators will also find lead. REMEMBER: you can search the same piece of foreshore again-and-again on a tidal river with a long history of use by man.

If you accept the advantages and disadvantages of discrimination outlined above but you still wish to search "money parks" and to use the "plastic bowl" search method on tidal riversides the question arises whether you should buy a single detector with both discrim. and non-discrim.

circuits, or two units – one discrim., the other non-discrim. The arguments for and against have so far failed to provide an answer for me, though I suspect – with nothing more than a "feeling" as evidence – that dual-purpose detectors never quite match the performance of units designed for one job.

A selection of metal objects found with a detector including: army badge, fishing weight, padlock, Victorian key, silver lighter and lead bullets.

Searches carried out downstream of old bridges always produce interesting finds.

Beachcombing

It is common knowledge in treasure hunting circles that beaches hold vast quantities of modern coins and jewellery. It is also common knowledge that beaches hold vast quantities of silver paper, and that coins and jewellery sink with alarming rapidity in sand – beyond the range of many detectors in a matter of hours, though they "re-appear" near the surface during and just after violent storms and very rough seas.

These then are the limiting factors to be borne in mind when planning a beach search. How can present-day metal detectors improve our success rate at the seaside? Well, so far as inexpensive BFO units are concerned nothing has changed during the past few years; they are still very poor producers of coins and rings on public beaches. They have rather a shallow depth range and only when one is able to work the sands as the coins and rings are dropped is it possible to achieve consistently rewarding results.

So far as IB machines go there have been great advances in depth penetration – but silver paper remains a problem, especially when working the high, dry sand at the top of the beach which often holds large numbers of rings. Certainly, the productivity of IB units when used on beaches has improved enormously – but best results are still gained by working in winter and early spring in stormy weather. Discriminators have largely eliminated the silver paper problem in the top six inches of sand – where most of the silver paper, but FEW of the coins and rings are found. For fast pick-up of modern cash lost very recently on beaches the best (i.e. deepest penetrating) discrims are certainly worth using. But I am convinced they are outperformed during wintertime searches by non-discrim. IB units because (a) silver paper is not much of a problem in winter when heavy seas wash it away quickly; (b) the greater depth penetration achieved with non-discrims is still helpful even though the waves will have "ploughed up" the sand and brought coins and rings nearer the surface.

And then there is pulse induction – with its extremely

deep penetration and its elimination of silver paper. I know a number of serious beach hunters who swear by their pulse machines and who can produce "oodles" of rings to back their enthusiasm. They claim – rightly – that the problem of poor pin-pointing does not matter greatly on a beach. It can be overcome by using a large plastic sieve. One simply gets a signal from the general area of the find and digs a deep shovelful of sand which is then thrown into the plastic sieve. A quick check with the detector reveals immediately whether or not the find is in the sieve. Coins and rings at twelve to twenty inches – far beyond the penetration of other units – are thus located by pulse users. However, I and other interested enquirers have noted that although the pulse men certainly get the gold rings and the coppers, they have very little success at finding silver coins and silver jewellery – the price paid for silver paper elimination because the detector sees most of these silver objects as junk silver paper. Now I'm not knocking gold rings – there's nothing better than a choice gold ring to warm the cockles of a treasure hunter's heart. . . . But the present cost of pulse machines, bearing in mind that it is only on beaches, and at the expense of silver coins, that they can be used for amateur treasure hunting activities, make them very specialized units to be considered only by those who intend to confine their activities to modern beach combing.

For the average treasure hunter equipped with a moderately priced IB unit (£50–£100 at present) the best search method (apart from working beaches in mid-winter) is to dig a pit in an area that has already produced a few good finds, and to work the bottom of the pit down to three feet. First clear off six inches of sand over an area approximately four yards long and one yard wide. Sweep this area with your IB unit very carefully because it is at this range – six inches down to sixteen inches – that many high-value modern rings will come if the beach has been well-used in recent weeks. After searching the entire area take up your shovel once more and dig out a further six inches of

41

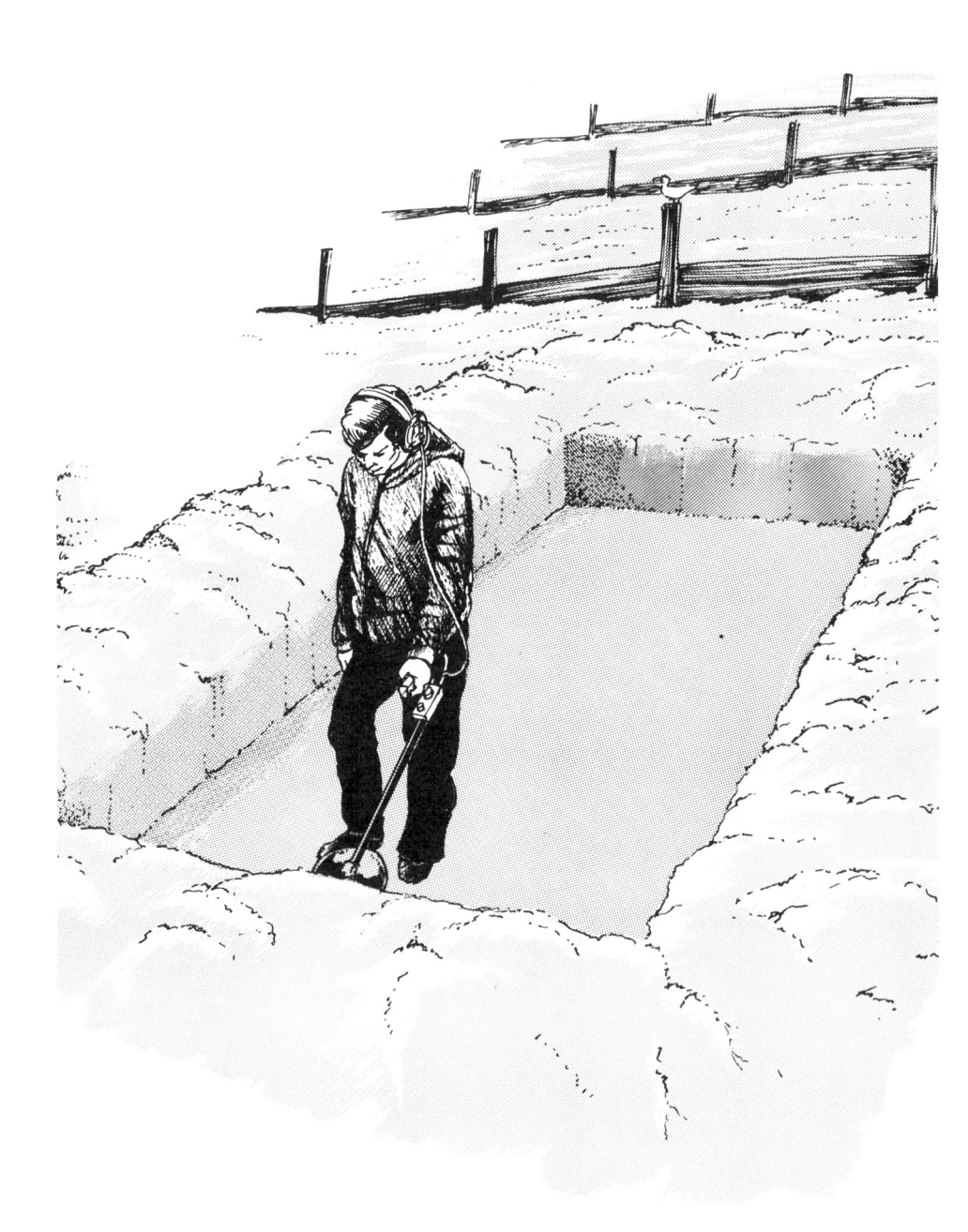

sand before trying the detector once again. At this depth – twelve to twenty two inches – you should begin to find Victorian coins and jewellery – IF the beach was used before 1900 by Victorian holidaymakers. A third layer should be removed if research has definitely indicated that the beach was indeed popular in the last century. On many beaches this will bring you within penetration range of the "hard-pack", the near-solid shelf on which the sands rest, and where, if you are very lucky, you could find a large accumulation of lost coins and jewellery which have sunk through the sand to that level.

Similar techniques – digging pits to a depth of three feet – also give good results when tried near sea-walls, groynes, and piers where sand has accumulated due to tide and wind movement. On pebble and shingle beaches the removal of two or three feet of material will produce good finds if PEOPLE have used the beach in the past. After working shingle patches on otherwise sandy beaches it can also pay to dig pits if the top ten inches of shingle have yielded coins and other finds. On rocky coasts with a history of shipwreck exposed ledges which have been undercut by tidal action should be raked when accessible and the detector passed over the material removed. Water immersibility helps when searching ledges and rockpools which do not dry out at low tide.

Here are two ideas I have seen employed by beach workers during the past few years which might profitably be tried on your local beach. Many of the oldest structures on our beaches – piers, groynes, etc. – are made from cast-iron and it is therefore impossible to search close to them with a detector. I recently saw a couple of treasure hunters working beneath the iron pier on a beach at a Yorkshire coastal resort. They had a large tarpaulin sheet of the type used by lorry drivers which they laid out flat on the sand. They then dug a trench along the line of the pier's "catwalk", throwing the sand onto the tarpaulin sheet as they worked and raking it flat. When the sheet was almost covered to a depth of four inches they searched it with their

One of the best search methods on the beach is to dig a pit and to work the bottom with your detector.

detectors to extract finds, then lifted the outer side so that the sand fell back into the trench. The sheet was then moved one length down the line of the pier and a new trench dug out. They told me it took a couple of days' hard digging to work right round the pier and that they did the job twice a year – in autumn at the end of the summer season, and in late spring following storms at the beginning of the year. Their haul on the day I watched them at work included several rings dug from depths of two to three feet. Strange though it may seem, they claimed their results were getting better each year, and that the amount of 19th century coinage found had remained consistently high during the three years they had used this method. Both treasure hunters owned non-discrim. IB units.

The second idea was thought up by an amateur treasure hunter who lives in a seaside resort and who spends most of his treasure hunting hours working the same stretch of sands. In order to work out how tidal movements affected coins and rings lost on the beach he devised a method for working out river movements. Several hundred marked discs of average coin and ring weights were dropped by the treasure hunter on the most popular parts of the beach during summertime. Over the next few months he carried out – in addition to his normal searches – a detailed survey of the beach, always working along the same lines sighted between the pier and various prominent buildings on the seafront. Whenever he recovered one of the marked discs he plotted its location, depth, and date of find onto a large-scale map of the foreshore and was thus able, during a full year's research, to gain very useful information about currents and coin movements over the local sands. He reports that his finds rate has increased greatly now that he knows the likely "dropping points", and that he is fast gaining a reputation as the most knowledgeable beachcomber in town. . . .

With reference to finds made on the "hard-pack", beginners who have yet to attempt deep digging on beaches should note that many objects located at this depth become encrusted with a material best described as "soft sandstone". It is the very heavily compressed sand on the bottom of the beach which, when dug up, comes out like pieces of broken biscuit. I have known gold sovereigns to be completely encased in this material – and very difficult to see. On some beaches – notably those on the east coast – streaks of iron-staining are also often associated with the "hard-pack". Unless one is aware of the problem, this can lead to valuable coins being discarded because the lump of sandstone in which the coin is encased can be mistaken for a piece of ironstone by users of non-discriminating detectors. You should always break open large lumps of material to which the detector has reacted.

Let me add another tip which can lead to handsome profits from beaches. From time-to-time (and most often in winter) reports are published in national newspapers about well-known seaside resorts which "lose" their sandy beaches. This phenomenon often occurs overnight when a violent storm or a sudden change in local currents sweeps away all the sand on a beach, carrying it out to sea or several miles down the coast. Such "happenings" attract the attention of newspaper reporters because the resort can no longer boast of its "miles of golden sands". All that is left is the "hard-pack" – when the oldest coins on the beach will be only an inch or two beneath the surface. Keep a look out for these valuable news items; when you see one DROP EVERYTHING and get down to that beach as fast as possible. . . .

Finally, a word to those beginners who, after trying their luck once on a beach, give up because they are frustrated by silver paper, ring-pulls, and bottle tops. The word is PERSEVERE. Stick at your beach search for a full day, place every junk item you find in a bag for disposal in the nearest litter bin, but keep searching. The only difference – apart from type of detector – between successful and unsuccessful beach treasure hunters is that the successful ones know that PATIENCE BRINGS PROFITS.

Victorian seaside resorts always provide interesting finds for treasure hunters.

Treasure Hunting Clubs

I strongly recommend membership of a treasure hunters club to ALL newcomers to the hobby. The benefits are enormous, and the small annual membership fee is rapidly re-couped from the extra finds located by those who make the effort to join a club. It matters little whether the club is a national organisation or simply a group of local enthusiasts who have banded together to pool resources; the advantages are much the same, apart from the fact that membership of a national club will provide the newcomer with useful information about areas of the country likely to be visited only during the holidays.

Perhaps the most important benefit to newcomers is the opportunity to meet more experienced enthusiasts who will pass on advice and opinions about treasure hunting equipment and techniques. You will, of course, hear all shades of opinion about the performances of metal detectors; but if you use what you hear in conjunction with what you see you will soon reach profitable conclusions about which detector you ought to buy next time you change models. This benefit alone could save you many times the annual membership fee.

By watching more experienced users at work with their detectors your own performance is sure to improve. You will soon get to know the most successful club members; they are usually the people who have least to say, but who simply get on with the task of making finds. They work methodically, do not dash around the site like race-horses, are experts at extracting objects from the ground without making a mess, and ALWAYS have the best collections of finds. Two or three weekends spent in their company are worth the entire contents of dozens of books on the subject.

If club members are pooling resources and research information, or if members receive site information as a benefit of membership, your finds rate is certain to increase. RESEARCH is the key to all rare and valuable finds – but if you are working alone it can take weeks, even months to locate suitable sites. One of the problems for the average newcomer is that weekends are the only time he or she can

Treasure hunters arriving for a day's outing by riverside, organised by the Treasure Hunters Club.

devote to the hobby. Unfortunately many of the best research sources – libraries, archives, museums, and newspaper offices – are closed for part or all of the weekend. But in a group of ten or twenty local enthusiasts there is usually at least one member who has spare time between Monday and Friday to devote to research.

Another money-saving benefit of co-operation between local enthusiasts is that the cost of travel to sites is greatly reduced by a car-sharing scheme. Petrol is now so expensive that shared transport is virtually essential when visiting sites more than fifty miles from home. If YOU are a car owner you will have no trouble filling spare seats with members who do not have their own transport and who will be delighted to share expenses with you.

Treecombing

Trees are to amateur treasure hunters what pointing skeletons were to Long John Silver and his shipmates; they indicate buried "loot" – lost coins, jewellery, badges, medals, and other metal relics from the past few hundred years. Learn to identify old trees and the clues to productive sites they provide and your "profits" will increase a thousandfold.

What do you look for when you go out with your detector to try your luck around "old trees"? Many people think of little more than height when trying to find old trees. "If it's tall it must be old" is the popular belief – but this is far from true. In fact the very oldest trees are more likely to be near-dead stumps, the remains of trees cut down by forest, park, and estate workers because old age made them dangerous. Yet time-and-again I have observed newcomers to amateur treasure hunting walk past an old stump without a second glance. So, for the benefit of those who want to improve their finds rate, let me set down a few useful facts about these most productive sites.

The oldest trees found in large numbers in Britain are yews, long associated with churchyards and also found growing in old estates and parks. YEWS CAN LIVE UP TO 1,000 YEARS. THINK ABOUT THIS. On a time scale it means that there are yews still growing today that were growing when William the Conqueror landed his troops from Normandy. They have been silent witnesses to the Black Death, the Wars of the Roses, the wars of all those Henrys and Edwards who battled so long with the irrepressible Welsh, the Civil War with its Roundhead and Cavalier skirmishes, the slaughter of the Scots who followed the Bonnie Prince, and the even greater slaughter of English, Scots, Welsh, and Irish in two world wars. What a story a single yew could tell if only it could talk. . . . Think of the bands of urchins who might have climbed its branches looking for birds' eggs; the numerous courting couples who stole kisses beneath them; the many families who have picnicked in their shade. . . . Then multiply all that by the number of yews that grow in our "treasure island".

Right: Coins held up by roots are much nearer the surface where there is no build-up of fallen leaves.

Below: In woods where there is dense vegetation a rapid build-up of fallen leaves occurs, making coins difficult to detect.

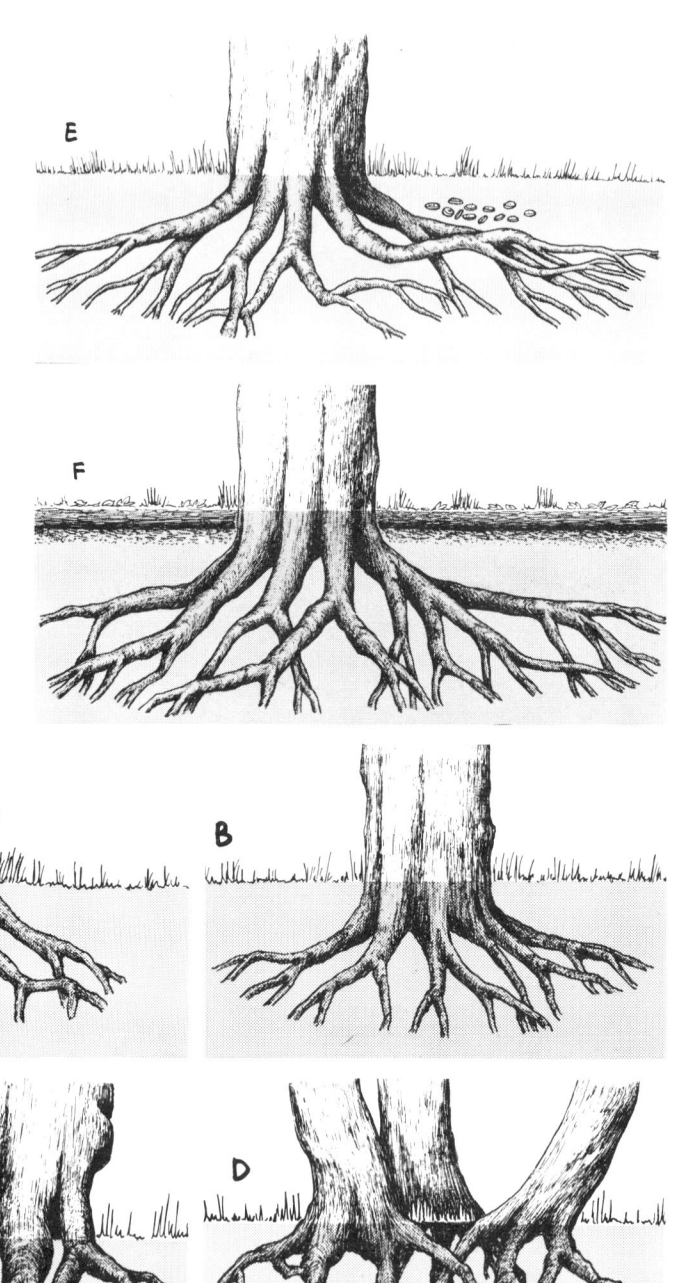

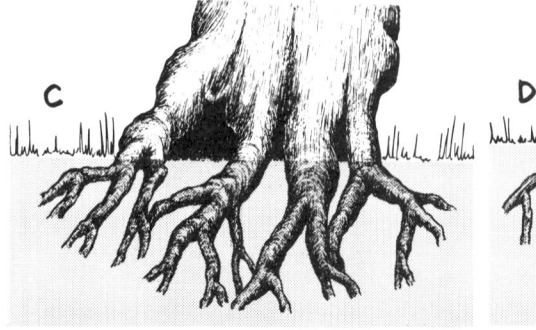

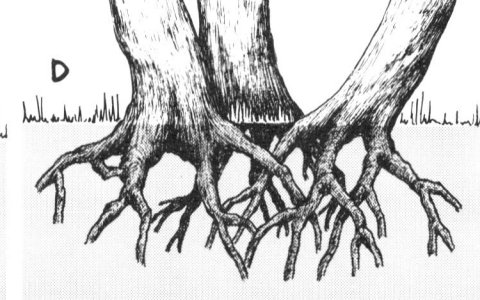

Opposite:
A. Tree $\frac{1}{2}$ m thick; too young to have old coins around it.
B. 1 m thick; 200 yrs. old.
C. 300+ yrs. old.
D. An ancient tree ring —500+ yrs. old.

Oaks come next in the longevity chart, though – surprising as it may seem – they live only half the number of years a yew can live. Still, five hundred years is a long time. A five-hundred-year-old oak has seen twenty six monarchs come and go – from the House of York to the House of Windsor – and because our population has increased dramatically during that time every old oak has probably been used by even more people for climbing, shade, and seclusion than has the oldest yew.

Ash, sycamore, Scots pine, and beech can all live up to three hundred years. Specimens old today were probably planted when Charles II sat on the throne; again, thousands of human beings will have made use of them down the centuries.

Some of these trees will have grown naturally – without the aid of man – in the vast forests that once covered parts of England – with remnants surviving in places such as Epping, the New Forest, Sherwood, and elsewhere. Others were planted (or encouraged to grow) by man for a number of reasons. First, as a source of timber for fuel and construction work. Ancient seaports depended almost entirely on local oak forests for shipbuilding materials; the remnants can be found around such old towns as Harwich, Southampton, and Dover. Secondly, as shelter for cattle and humans. If, while out walking in hilly countryside, you spot a belt of very old trees on the windward side of a farmhouse (or small hamlet) you have a site worth searching; the trees (or their earlier parents) were originally planted as a shelterbelt to provide protection from the wind. Thirdly, as ornamentation – in the geometrically planned parks and estates laid out in the 18th century by rich landowners, and (later) by public benefactors who gave many cities and towns their first public parks. Both yews and oaks were also planted in the distant past for religious reasons; tree worship was part of the Druid religion and Christianity absorbed the fetish when yews were planted near churches.

WHERE SHOULD YOU LOOK FOR OLD TREES WHICH MIGHT HAVE VALUABLE COINS AND

Many interesting finds have been made in the past by those who took the trouble to search inside the hollow stems of ancient trees.

OTHER RELICS IN THE GROUND AROUND THEM? Firstly, around old churches. I am not suggesting you should go into churchyards to use your detector; few people will have sat in a churchyard. But there are many places where churches have fallen out of use years ago – perhaps even been demolished. Look for old yews nearby and search around them if there is public access.

Secondly, around old farmhouses where the trees were

originally planted to provide shelter. Obtain the farmer's permission *before* you start the search. You will be amazed at the response you get when you tell the farmer you are very interested in old trees and their association with local history. Even farmers who rarely give permission for searches in open fields will almost always agree to a search around old trees. They often scoff at the idea of coins being found on ploughed fields, yet they seem to know instinctively that the trees on their land have a long association with the farm and those who have worked on it in the past.

Thirdly, around old village greens and in shelterbelts associated with ancient hamlets. These sites are especially productive where there are village traditions of maypoles, well-dressing, and other old customs.

Fourthly, on roadsides – especially along roads built during this century which have cut through old woods and copses. You can spot sites like this by looking for belts of old trees dotted around both sides of a stretch of road. Do not confuse these productive sites with those regularly spaced single trees planted along roadsides either at the time the road was built, or more recently as part of landscaping schemes. The remains of old woods that have been sliced by a modern road always look untidy – often with half-dead trees, fallen trunks, and stumps cluttering the floor of the wood.

Fifthly, in *old* estates and parks – if you can obtain permission for a search.

HOW DO YOU IDENTIFY THE OLDEST TREES?

Firstly, by girth – i.e., the thickness of the trunk. A tree (yew, oak, ash, sycamore, Scots pine, beech) with a thickness of more than one metre will be at least two hundred years old. This means that in 1877 it was thick enough to attract people. On the other hand, a tree that is only half a metre in thickness today was a mere sapling a hundred years ago – insignificant as a tree for shelter, climbing, etc.

Secondly, by its state of health. A dead stump measuring one or two metres marks the site of an ancient tree – three hundred and more years old. If it was accessible to people in the past it could have large numbers of Victorian and earlier coins scattered up to several yards around it. A living tree which has dying branches and great holes in its trunk is called a "decrepid" – and could be as old, or older than a dead stump. Search around it – and in any large holes in the trunk which might have provided hiding places for hoards.

Thirdly, by the condition of its roots. As a tree gets older its spreading roots take up all the nutrients and moisture in the surrounding soil. This results in soil shrinkage and exposure of the roots. The oldest trees will have roots that seem to be "pushing" the tree out of the ground. Search in and around them.

Fourthly, by looking for ancient rings of trees. These occur when the main stem of a very old tree has died and three or four new stems have grown from around the old roots. The new "trees" (all part of the same tree) appear to grow in a neatly-planted ring. This is the only time you should pay careful attention to trees less than one metre in thickness; the central stem (now dead) was probably two or three metres wide before it withered.

Fifthly, by looking for "pollarded" oaks. These are always very old trees which were used in the past to provide large numbers of thin stakes for fencing or firewood. You can spot them by their thick stems topped by a "bush" of very thin shoots. Pollarding died out many years ago – but the trees survive and invariably provide productive sites because many coins and other objects were dropped by the men and boys who climbed them to cut the branches in the past.

HOW CAN YOU IMPROVE YOUR "FINDS RATE" AROUND OLD TREES? Firstly, by using a detector with good depth penetration – especially when you plan to search woods where an accumulation of fallen leaves will have covered the floor of the wood to a considerable depth. It is this "carpet" of dropped material which prevents those who use detectors with poor depth penetration finding old coins in woods.

Secondly, by selecting your trees carefully. When a very

Below left: The sites of old stumps often indicate a spot where a very old tree – perhaps 500 and more years old – has been felled.

Below right: A pollarded oak. The upper branches have been cut back many times to encourage the growth of thin shoots for fencing and firewood.

old tree grows in an open area, with little protection from surrounding trees and other vegetation, the leaves it sheds over the years are blown away by the wind. This greatly reduces the build-up of material around the base of the tree; lost coins are not so deep under these conditions. If, in addition the roots are very near the surface they will "hold up" objects dropped on top of them. A further benefit in shallowness of buried coins is gained when the soil contains large numbers of small stones and pebbles which also help to keep coins near the surface. Clay soils often contain these stones and pebbles.

Striking it Rich

It is highly significant that the most recent hoard find by a treasure hunter – reported to me just a few hours before I sat down to write this section – was located after several months of painstaking research and detective work by the man who set out to search for it. I'm sure the full story of this find will be widely publicised in treasure hunting circles before this book goes on sale, so I will relate only the bare facts here. The finder, who lives in Carlisle, was browsing one day through an old book on the history of Cumberland when he came across a reference to the discovery by a farm labourer in the 1850s of a pot containing 11,000 Roman silver coins. Only an approximate location of the hoard site was given in the old book so the treasure hunter next consulted modern maps of the area. Unfortunately the modern maps did not contain references to the find, though they did

Items found in Kent with a detector, including part of the helmet of the 6th Regiment Afoot (c. 1790).

show that the general area was still farmland. After looking at several older maps and talking to local people who had lived in the district for many years the treasure hunter was able to narrow the location down to a single field. Further proof of the field's potential as "treasure ground" was provided by the farmer who now owns the field; he told the searcher that the previous owner had found "a golden jug" while ploughing the field several years ago. After obtaining the present owner's permission AND reaching an agreement with him to share profits the treasure hunter, accompanied by his thirteen-year-old son and a friend, began a methodical survey of the field with their detectors. Within less than an hour of commencement the thirteen-year-old had found the first of several hundred Roman bronze coins so far recovered from what seems to be a scattered hoard. The search continues as I write. . . .

More and more stories like this are now being told by successful treasure hunters who have learned that time and effort invested in RESEARCH can pay handsome profits. So far as I am concerned, the most important fact in this latest success story is THAT A HOARD HAD ALREADY BEEN FOUND IN THE SAME FIELD. I was convinced long before the tale reached my ears that the best place to look for a hoard of buried treasure is a place where buried treasure has already been found. The evidence to back this theory is overwhelming. Time-and-again ploughmen, building-trade workers, civil engineers, and (more recently) treasure hunters have found hoards in fields where earlier, accidental finds had already been made. Few, if any, of the finders before the days of metal detectors were aware that their find was the second to be made. Perhaps they found out later when museum officials checked their records, but this information would have been of little use to them without a detector – unless they were planning to dig up the entire field. . . .

Things are different nowadays – yet, despite the "magic wand" given to us by modern electronics, the number of hoards found by amateur treasure hunters is still very small indeed compared to the thousands discovered accidentally in the days before metal detectors became available. I recall saying a few years ago that the reason for this lack of success at hoard finding was that there were not enough metal detectors in use. That *may* have been true three or four years ago, but it is no longer the case. The ONLY reason which can possibly account for our lack of success is that insufficient research on potential hoard sites is carried out by detector owners. I accept that research is a time-consuming business, and that many treasure hunters are unable to visit research sources because they have jobs to hold down Monday to Friday. Nevertheless, it still seems odd to me that those thousands of treasure hunters who have each spent £100 and more on their treasure hunting equipment should be content to hunt single coins when so many pots full of 'em await discovery.

To those people who doubt that hoards come in pairs I say: CHECK THE RECORDS. They show irrefutably that the practice of burying two (or more) pots was widespread in the past, especially during the Roman period. One of the reasons for the practice was that it provided insurance against total loss; if robbers found one of the pots the owner still had the contents of the second to fall back upon. Another reason for it was simply that it made account-keeping easier. This applied more to the larger hoards where the owner kept all bronze coins in one pot and all silver in another. If he was very rich he might also have kept gold in a third. It is also possible that this division of different metals, together with the positions the pots were buried in relation to each other, had some religious significance. Certainly, many hoards were buried in proximity to religious monuments, though the monuments (barrows, standing stones, etc.) may simply have been used as fixed marks from which to pace off the distance to each pot.

It follows from the above that if a treasure hunter could pin-point the exact location of every hoard dug up in his county in the past he would stand an excellent chance of locating his own hoard. Unfortunately it is impossible to

A selection of Roman coins found with a detector, from agricultural areas.

compile so comprehensive a survey. Very few hoard locations were accurately recorded before 1800; indeed, many were not recorded at all – they survive only as folklore. But records have been more accurately maintained in the 20th century and the locations of hoards found nowadays can be pin-pointed to within a few feet. To obtain this information requires little more than patient research. ALL local history books must be read from cover-to-cover; ALL archaeological reports must be scanned page-by-page; ALL local newspapers must be perused inside-and-out. Do this and – if you plot your information on a county map – you should eventually compile a hoard map of your county showing at least fifty locations where pots of treasure have been found in the past.

If possible, every location should be subjected to a methodical lines-and-pins survey with a detector capable of locating a pot two or three feet beneath the surface. (Permission for each search AND an agreement with each landowner to share profits would, of course, be obtained beforehand.) But if lack of spare time obliges you to concentrate on only a few of the sites you should select those where pots containing ALL silver or ALL bronze coins were found because it is on these sites that a second pot is more likely to await discovery. Question the landowner and his workmen very closely; if they have found even a single coin of a different metal in the same field you are almost certainly on a site where a hoard lies beneath the ground.

If time permits you should, during your research, note the locations of all single coin finds made in the past. A single coin – perhaps pushed towards the surface by earthworm action or the touch of a ploughshare – might have come from a buried hoard. Many farmers keep odd coins that turn up during ploughing. If you have the opportunity to examine them and they all date from the same period, or are all in a similar state of preservation then, again, the chances are that a pot lies buried in the field.

Ploughing and the passage of time will almost certainly have obliterated the original landmarks used by the hoarder who buried the pots, but a little thought and one or two questions can often narrow down the search area. Field names (e.g., "Barrow Field" or "Standing Stone Acre") can indicate that you are searching in the right spot. A farmer's comments on the condition of the land can be useful. A patch of deep, fertile soil used by a 20th century farmer would also have been used by a 3rd century farmer – who would have favoured such a location for the burial of pots because cultivation of the topsoil would hide evidence of burial. It is far more difficult to disguise the evidence of deep hole-digging in infertile, stony ground where the digging of a hole would be as plain as a pimple on a nose. . . .

Of course, research is not solely about hoard hunting; the serious researcher also records during his browsing through books and newspapers details about local customs, fairground sites, toboggan runs, the building of bridges, the locations of river fords, and the places where, a hundred and more years ago, people gathered for sports, picnics, and outings. He also compares old and modern maps to ascertain where changes have taken place – in the pattern of woods and forests, the courses of rivers and streams, and the local coastline. Nor is research solely about reading dusty old books, newspapers, and maps. PEOPLE, es-

Georgian buckles, cufflinks, buttons etc., found with a detector.

pecially old people with good memories, can provide some of the best locations for coin and relic hunting. The dedicated researcher practices and masters the art of "pumping" these old folk for clues. He LISTENS when they talk of the "good old days"; he RECORDS what they have to say about half-forgotten local customs; he ACTS on the knowledge gained. . . .

Group of children proudly displaying their finds on a tip (c.1900).

Dump Digging

If YOU want high profit IN MONEY from your amateur hunting then, unless you can find a hoard, you will have to turn your attention to dump digging.

There are still almost 50,000 undug refuse dumps in the British Isles – council dumps, mansion dumps, farm dumps, and *hoards* of old bottles gathering dust in cellars and sheds. Many are worth substantial sums of money: a cobalt-blue, marble-stoppered Codd bottle made around 1890 is now worth at least £150 in fine condition; a bear's grease pot lid with a picture of a bear could be worth £100; even humble ginger beer bottles fetch £10 each if they have coloured tops and attractive trade marks. Find ONE dump and you could unearth ONE THOUSAND BOTTLES from the site during a few weeks of hard digging.

As for the market, it is INTERNATIONAL. Americans by-the-million collect old bottles; so do Australians, New Zealanders, South Africans, Canadians – not to mention the large number of Britons who have also been bitten by the bottle bug. ALL OF THEM WANT TO BUY THE TYPE OF BOTTLE YOU CAN DIG UP ON BRITISH DUMPS CONTAINING PRE 1900 REFUSE.

How to find sites: In many towns it is possible to pinpoint the location of refuse dumps used during the late Victorian period by consulting 19th century records. Public libraries, town or city archives and the offices of borough council refuse disposal departments often keep copies of old minutes and reports which give the exact locations of dumps used by the town before 1900. A search for such documentary information should always be your starting point when attempting to locate an old town dump. Even if the 19th century records have been lost or destroyed you will be able to eliminate many 20th century sites by a careful check on more recent documents which can be consulted at such places. Obtain an Ordnance Survey map of the district and mark on it the locations of all post-Victorian dumps. Their positions might indicate likely hunting grounds for earlier sites; if a town dumped its refuse on low-lying marshes or in worked-out quarries thirty years ago it

Black and white pot-lids dug on various tips. For list of prices see – "Collecting Pot-Lids Price Guide" by A. Ball.

is quite probable that similar sites were used when organized collection and disposal first started.

Geographical clues to the whereabouts of possible sites can also be gleaned by comparing Ordnance Survey maps drawn a hundred years ago with modern maps of the same area. Clay and chalk pits which appear on early maps but which are not now shown are quite likely to have been filled with refuse; changes in the level of land alongside river-

Even 17th Century Bellarmine jugs, like this one have been found on Victorian rubbish tips.

Black glass embossed beer bottle.

Internal screw-top, lemonade and beer bottles.

Transfer printed ginger bottles.

66

banks can also indicate earlier dumping; and the sites of brickworks which ceased manufacture sixty or more years ago might also be shown on the early map. All such sites should be marked on your modern map as worthy of close inspection.

In cities and larger towns, where hundreds of thousands of tons of refuse were disposed of every year between 1860 and 1900, many dumps would have been used and it is most unlikely that the exact location of each site will be recorded in old city and borough records. Often the minutes merely report the appointment of a certain private contractor who undertook to dispose of refuse for a number of years at an agreed figure per ton. In such cases other lines of enquiry must be followed. A check on Victorian trades directories at the local library might reveal that this contractor owned a fleet of barges or horse-drawn waggons. The contractor, or his successors, could still be in business today and a polite letter to the company secretary often produces evidence of where the company's old dumping grounds were sited. On more than one occasion where I have followed up such a report I have been put in touch with a retired employee who actually worked for a Victorian contractor. Needless to say, such men are mines of information.

Whatever line of enquiry you follow when checking written information on early dumps you must eventually

Selection of miniature bottles often found in tips.

Early Victorian Stoneware figural flask.

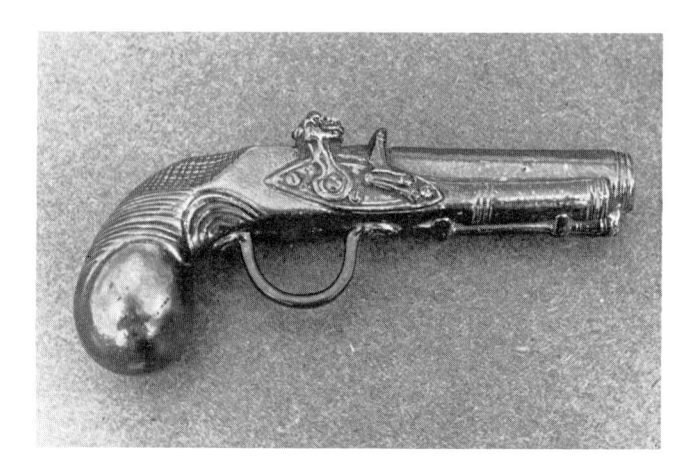

Late 19th century coloured medicine bottles, left to right: emerald green "FISHERS SEAWEED EXTRACT"; cobalt blue "PRICES PATENT CANDLE"; red amber "Wm R. ADAMS MICROBE KILLER". For prices see "BOTTLE COLLECTING PRICE GUIDE" by Gordon Litherland.

A family of bottle diggers selling their finds at the Burton-on-Trent Bottle Show.

confirm your findings by visiting the sites. Often a physical search of marshes, old quarries, riverbanks and canal sides is the only method of locating a site which can be used because old records no longer exist. It is a search method I wholeheartedly recommend; if carried out thoroughly it *must* lead to the finding of a worthwhile site.

Many newcomers to Victorian dump-digging miss excellent sites because they neglect to search areas of countryside which cannot conveniently be reached by car. The only sites you are likely to find if you refuse to venture more than a few hundred yards from your parked vehicle are those which have already been found by many other bottle or relic hunters, or those on which 20th century rubbish has also been dumped. The best sites lie well off the beaten tracks – on silted creeks and along the backwaters of river estuaries, on lonely marshes, in deserted quarries and alongside derelict canals. There are few roads to get you there and a boat would probably run aground. The only solution is to sling your digging tools across your back and reach the site on two legs. The reward for such effort is likely to be a choice site into which no other digger has put a fork.

Rockhounding

There is nothing difficult about finding semi-precious gems and valuable fossils in Britain. Although we don't have diamond mines and huge dinosaur bones in these islands, we do have a vast assortment of commoner gemstones such as amethyst, rock crystal, agate, jasper, and fluorite, together with numerous fossils such as ammonite which can be worth several pounds apiece.

The equipment needed to hunt both semi-precious gems and fossils include geological maps (obtainable from HMSO), hammers, and an assortment of steel chisels. Add to these some warm clothing and a good pair of boots and you will be fully kitted for your first hunt.

Gemstones are found in those regions where volcanic activity occurred in the distant past. Such areas are identified by their granite rocks – Cornwall, parts of Wales, the Lake District and Northern Scotland being obvious examples. Because metals such as lead, tin and copper also result from volcanic activity you need only concentrate your search around abandoned mines in order to ensure success.

Fossils, on the other hand, are associated with sedimentary rocks – i.e. areas where seas and lakes once flowed and where their muddy and sandy bottoms have been transformed over millions of years to shale, slate and various types of sandstone. Limestone regions are also good for fossil collecting because limestone is made from the skeletons of long-dead sea-creatures. Classic fossil-hunting areas can also be found in many other locations. At your local public library you will be able to consult geological reference books which list the exact locations of *all* fossil producing areas in the British Isles. I recommend coastal sites as ideal to learn more about fossil-hunting because many of the specimens to be found on beaches have already been loosened from the surrounding rock and shale by the pounding of waves. By collecting in this way you will not risk damaging your finds when chipping them out of the parent rock.

It is of the utmost importance that all readers are fully aware of the dangers associated with old mines and with

*WHERE TO HUNT FOR
SEMI-PRECIOUS STONES*

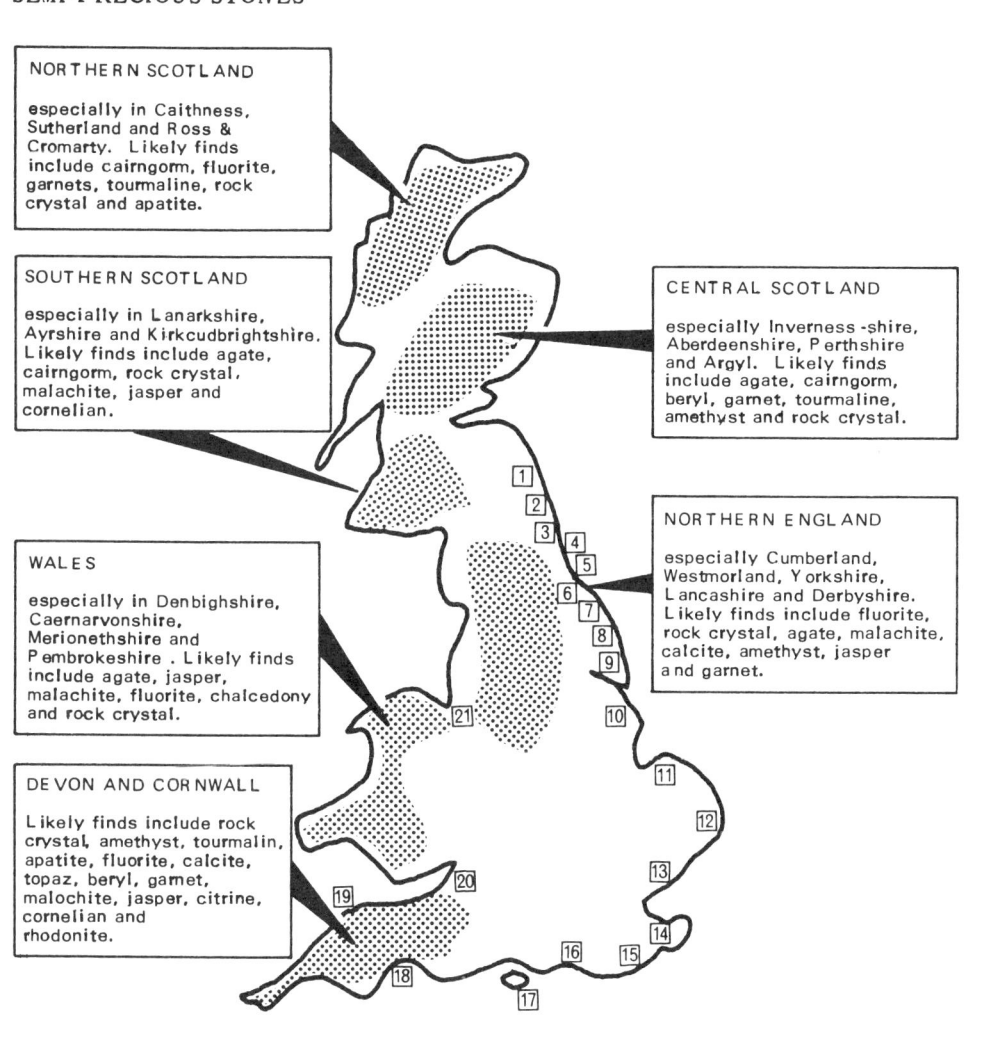

NORTHERN SCOTLAND

especially in Caithness, Sutherland and Ross & Cromarty. Likely finds include cairngorm, fluorite, garnets, tourmaline, rock crystal and apatite.

SOUTHERN SCOTLAND

especially in Lanarkshire, Ayrshire and Kirkcudbrightshire. Likely finds include agate, cairngorm, rock crystal, malachite, jasper and cornelian.

CENTRAL SCOTLAND

especially Inverness-shire, Aberdeenshire, Perthshire and Argyl. Likely finds include agate, cairngorm, beryl, garnet, tourmaline, amethyst and rock crystal.

WALES

especially in Denbighshire, Caernarvonshire, Merionethshire and Pembrokeshire. Likely finds include agate, jasper, malachite, fluorite, chalcedony and rock crystal.

NORTHERN ENGLAND

especially Cumberland, Westmorland, Yorkshire, Lancashire and Derbyshire. Likely finds include fluorite, rock crystal, agate, malachite, calcite, amethyst, jasper and garnet.

DEVON AND CORNWALL

Likely finds include rock crystal, amethyst, tourmalin, apatite, fluorite, calcite, topaz, beryl, garnet, malochite, jasper, citrine, cornelian and rhodonite.

SOME FOSSIL COLLECTING AREAS AROUND THE ENGLISH COASTLINE

1. Berwick on Tweed
2. Cullercoats
3. Hartlepool
4. Redcar
5. Kettleness
6. Whitby
7. Scarborough
8. Bridlington
9. Hornsea
10. Grimsby
11. Wells
12. Southwold
13. Southend
14. Herne Bay
15. Hastings
16. Worthing
17. Shanklin
18. Budliegh Salterton
19. Barnstaple
20. Weston-super-Mare
21. Flint

sheer cliffs on exposed coasts. A long-abandoned lead mine on a desolate moor is the ideal spot to collect fluorite specimens; it is also a very dangerous place for those who run about carelessly and do not keep alert for hidden mine shafts. Similarly, anyone foolish enough to wander along the base of sheer cliffs without first checking on the number of hours before the tide rushes in only has himself to blame if he has to swim for his life. You have been warned.

The best way to identify semi-precious stones and fossils found during your expeditions is to take them to museums in the areas where you find them and compare them with exhibits on display. This is a very useful exercise because it will also help you to become familiar with other specimens before you find them. Look carefully at the information written on the identification cards in museum display cabinets; they often give exact locations where spectacular finds have been made in the past.

Searching for Gold and Pearls

Surprising though it may seem to some readers, it is possible to find gold and pearls in the British Isles.

In the mid-19th century a "mini-Klondike" occurred in the Helmsdale region of Sutherland when the Duke of Sutherland engaged a family of tinkers to pan for gold in the streams of that valley. The tinkers, who knew absolutely nothing about gold recovery and whose only tools were home-made frying pans, successfully recovered gold worth £50,000.

Many geologists are convinced that a vast amount of gold still is beneath the rocks of Sutherland – perhaps as much as there is beneath the ground in South Africa. Unfortunately those thick veins – if they exist – will be thousands of feet deep and the cost of mining them would be astronomic. Nevertheless, it is still possible to pan for gold in Sutherland, as it has been proved by several amateur treasure hunters during the past few years.

Further south, in Lanarkshire's Leadhills area, gold panning was – until the beginning of the century – a part time occupation for most of the local leadminers. Whenever a local girl became engaged to be married all the men in the vicinity turned out and panned local streams until they had found sufficient gold to make the girl's wedding ring. Now that leadmining has ceased in the area few people work the local streams. But gold *is* still waiting to be found there.

In England and Wales some successful gold panning has been carried out in recent years in Cornwall and North Wales, but it is in Ireland around the Wicklow Hills that the richest finds have been made. There are families living in that area who make regular trips to London to sell the gold they pan from streams on their farms. Just how much they find every year is never disclosed, but this is another area of the British Isles thought by geologists to be immensely rich in the yellow metal.

There is only one way to learn the art of panning for gold: buy yourself a gold pan and try your luck, preferably in one of the above mentioned areas. The skill of swirling the pan in just the right way so that the flakes of gold don't run off will come to you – just like learning to ride a bike.

Pearls are easier to find than gold – if you know where to seek them. They are found inside the shells of freshwater mussels which live in all unpolluted rivers in Scotland, Northern England, North Wales and Ireland. If a river in any of these areas has salmon or trout then it will have freshwater pearl mussels. To catch your mussels you need a pair of waders, a glass-bottomed bucket, and a forked stick. Find a stretch of river with a sandy bottom, wade in up to your waist, and use your glass-bottomed bucket to scan the bottom seeking the black shapes of the mussels – which look rather like giant versions of the mussels found on rocky coastlines. Hook them from the water with your stick. If the mussel contains a pearl the outer shell will be oddly marked. You will have to open one or two before you find a specimen containing a pearl; when you do keep the shell so that you can recognise pearl bearing shells in future. All specimens without these markings should be returned to the river. The pearls you find will be blue, green, brown or white. The white specimens, if perfectly round, are worth from £1 to £1,000 – depending on size.

The Right Approach

If research is the key to treasure hunting success, then permission is the lock which must be opened if you are to have your success. The trouble with my clumsy analogy is that finding the key is simple, but getting your hands on the lock often proves impossible.

Why is it that amateur treasure hunters find it so difficult to obtain permission to search sites which research has shown *must* hold valuable coins and relics? The answer in a word is: IGNORANCE. Ignorance on the part of landowners about what amateur treasure hunting is and what metal detectors do; and ignorance on the part of amateur treasure hunters about how best to approach landowners when seeking permission.

Can you imagine the amount of money that could be recovered from Hyde Park by a small squad of experienced detector users? Enough – if the treasure hunters gave all the modern money to the Parks Dept. – to pay for the upkeep of the park for a year, paying the wages of all the staff, *and* the cost of flowers, trees, and shrubs. The treasure hunters would carry out the search for nothing more than the privilege of keeping all the pre-decimal coinage and a percentage of the unclaimed jewellery. Think of the charities that could benefit if searches like this were carried out in ALL major parks and the modern cash given to them.

Of course, the immediate reaction of the bureaucrats with the power to bring such schemes to reality would be to cackle like broody hens about damage to the grass and to reject the idea with visions of bomb craters, picks, shovels, and J.C.B.'s swimming before their eyes. This is the price we treasure hunters have to pay for our "public image" as looters and desecrators; an image we have gained thanks to the activities of a tiny minority of mindless goons who have gone into public parks equipped with picks and shovels and made disgusting messes. Unfortunately the goons also happened to be carrying metal detectors – and the inevitable result has been a total ban on all detector owners who might otherwise have enjoyed the pleasures and rewards of searching these most productive sites.

If only detector manufacturers and dealers could apply some sort of intelligence test to anyone wanting to buy a detector; or fit explosive devices into the heads of metal detectors that would trigger if the owner dug a single unsightly hole in the ground. ...

In the absence of a national scheme to clean up our tarnished image, I believe the answer lies with each individual who must, in his or her small way, prove to the local community that responsibility and good sense are qualities that need not fly out of the window when one becomes an amateur treasure hunter. Let me explain how YOU can do a good "public relations" job for the hobby AND obtain permission to search some profitable sites.

First, you require some photographs showing yourself at work with your detector. Enlist the help of a friend or relative if you are not a camera user and make sure the photographer gets a series of shots – perhaps taken on your front lawn – of you extracting a coin from a patch of grass. The photographs must be sharp and clear; they must also show a neat cap of turf being cut, the coin extracted, and the cap of turf replaced in such a way that the spot from which you removed the coin is indistinguishable from the surrounding turf. It should not be difficult to obtain these photographs because what they will show is the way you work at all times when out with your detector.

Having obtained several sets of prints, which should be numbered in sequence, you must now sit down and compose a letter which will accompany the photographs, and which you will send to the Superintendent of the park you hope to search. Here is an example of the sort of letter required:

> *Dear Sir,*
> *As a local amateur historian I would very much like to carry out a search in the grounds of .... Park in order to recover, with the aid of the electronic device shown in the enclosed photographs, some of*

These sketches show how a coin is extracted from grass-covered ground. Make sure the photos you send to the park superintendent show you doing the same. Use a sharp, broad-bladed sheath-knife for best results. Leave part of the circle uncut to act as a hinge when you replace the cap.

Probe the exposed soil with the knifeblade until you locate the find. Keep the soil in the hole.

There should be no evidence of your handiwork once the cap has been replaced and pressed down with the heel of your shoe.

the items lost in the park during the one hundred or so years it has been used by citizens of our town. As you can see from the photographs which were taken on my front lawn, recovery involves nothing more than cutting a neat cap in the turf which is replaced immediately after removal of the metal object. It does not make a mess and the surface of the ground remains unchanged on completion of the operation.

The items I would like to recover include coins, metal buttons, badges, etc., all lying in the soil at a depth of less than ten inches. The electronic device will also locate pieces of iron and similar "junk" which can, as you know, cause damage to grass-cutting equipment. I will gladly remove these hazards to your equipment if you give permission for the search, and I would also be pleased to use the instrument to locate buried pipes, manhole covers, lost tools, or any other items you might be interested in finding.

If you would like a demonstration of the instrument before reaching a decision I will gladly come along to the Park at any time convenient to you so that you can see it in use. I enclose a stamped addressed envelope for your reply and remain,
                                    Yours faithfully, . . .

If as a result of this letter you are granted an "audition" by the Superintendent make a special effort to keep all your "extractions" as neat as possible. Take with you a wide-bladed sheath knife with a good sharp edge. When you cut caps in the turf leave small sections uncut to act as "hinges". This ensures that the caps are replaced in exactly the right positions. Once they are pressed down firmly with your foot even you will have difficulty in identifying the spots you have worked.

Be prepared to answer any questions put to you by parks officials, and to accept a compromise if you are given permission to search only roughly grassed areas. If the question of damage already done by "mindless goons" is raised suggest that you might act as a semi-official "watchdog" and keep unauthorised searchers out of the park grounds.

It is worth bending over backwards to obtain even the most limited official approval for this first operation. Once you can produce a letter of approval from one parks department it becomes an "open sesame" to more. Each successful application improves your chances of even greater rewards.... Today Blogg's Recreation Ground; tomorrow Buckingham Palace Gardens.... And don't forget to take lots of photographs of the "before-and-after" variety showing stretches of park grass on which you have worked. Build up a folder of past successes in this way and it could become your passport to almost any park in the land. ...

I have already mentioned in the section on trees that farmers do not take unkindly to requests for pemission to search around old trees. Try extending such permission – AFTER you have done a good job on the trees – to the farm fields, especially if you have research notes that record the finding of coins in the vicinity. As with park keepers, the problem with farmers is that they imagine you digging enormous craters in their cow pastures, so make sure the farmer WATCHES you at work around the trees. Point out to him that your detector penetrates no deeper than his plough, and that you could be of service to him (free of charge) in locating underground pipes or removing dangerous hunks of scrap metal from his land.

Letters on the lines suggested above can also be sent to private landowners in your district; also to the owners of large houses with gardens likely to hold coins and relics.

IT IS OF THE UTMOST IMPORTANCE THAT AS AN AMATEUR TREASURE HUNTER YOU ARE

AWARE THAT AT ANY MOMENT DURING YOUR SEARCHES YOU COULD STUMBLE UPON A HOARD OF GOLD OR SILVER. Keep this thought in mind constantly when seeking permission for searches from farmers and private landowners, and reach agreement with the owner BEFORE THE SEARCH COMMENCES to share the profits from any hoard you locate.

Ideally the agreement should be written down and signed by you and the landowner, but it is often impossible to persuade landowners to sign pieces of paper simply because they are quite convinced you will find nothing of value.

If the landowner IS attuned to the idea of valuables hidden on his land it is in the interests of BOTH parties to sign such an agreement. YOURS because without an agreement you could lose a share of a bronze hoard; HIS because without an agreement he could lose a share of a gold or silver hoard. Also, keep up-to-date on the administration of the law concerning Treasure Trove. There have been some very odd decisions by coroners recently – including a case where an amateur treasure hunter lost a reward of £15–30,000 because "an expert" at the inquest stated that the "bronze" coins the treasure hunter THOUGHT he had found (and therefore did not declare) contained "between one and nineteen percent silver". On this piece of "expert" evidence the coroner confiscated the hoard. With this case in mind, I urge any reader who finds a hoard – even if every coin catalogue in the land states that the coins are bronze – to take them to the local police station and hand them over pending the outcome of further deliberations by the local coroner, the local museum, Old Uncle Tom Cobbley, and All on whether or not the coins are bronze and, therefore, NOT subject to the law of Treasure Trove.

PERMISSION – the lock on the gateway to treasure hunting success – can bring you the right to search anywhere. There are, of course, many places where permission will NEVER be granted for your activities – including scheduled ancient monuments, military installations, and the back garden of the house owned by your County Archaeologist. Indeed, unless he is a very enlightened gentleman and not at all like the archaeologists I have met, your County Archaeologist will deny you the right to search your own back garden. But he can't prevent you from searching open fields and gardens if the owner has given permission – unless he obtains a preservation order first. . . .

I'm afraid you must be prepared for this sort of treatment from many archaeologists, but there are one or two enlightened archaeological societies that have recently conceded to amateur treasure hunters that metal detector searches carried out in unstratified plough-soil do not damage archaeological sites and can even *help* archaeologists in putting together the jigsaw of history. One of the best examples of this enlightened approach to amateur treasure hunters is revealed in a pamphlet published by the Archaeology Department of Norfolk Museum Service and the Norfolk Archaeology Unit. It represents such an important breakthrough in relations between amateur treasure hunters and archaeologists that it is worth re-printing in full here:

"*Most archaeological finds are discovered by people working on, or walking over the fields. Archaeologists are always delighted to examine and identify any finds: do not be afraid that you have only found rubbish. The shapeless scraps of metal which your detector picks up, and the fragments of crude pottery which you see while you are scanning the ground are important to archaeologists. Such finds can turn up anywhere, in ploughed fields, on beaches, on commons, or in gardens. In ploughed fields most detectors will only indicate objects in the plough-soil. If you dig a bronze brooch from the plough-soil you will not be damaging any archaeological layers that there might be below. But that brooch is evidence: it may be the first Saxon brooch from what is thought to be a purely Roman site. You may not know the difference at first, but an archaeologist* will and will be pleased to explain it to you, and your discovery might easily alter our ideas about an area. If no archaeologist ever sees your finds, or if you do not record which field you found them in, valuable information is lost, so please record your finds and let us see them, and help us to solve some of the problems of Norfolk's past. If you find a large metal object, it might be below the plough-soil. It could be buried in a pit, or below the floor of a building: a team of archaeologists takes several days to uncover an object like this and to study its position in the ground before digging it out. Ideally archaeologists would like to excavate objects from below the plough-soil themselves, so please try to contact an archaeologist first. This is not always possible, so if you do dig up an object like this, please try not to disturb the surrounding soil; if necessary archaeologists can then go back to the spot and try to discover why the object was put there in the first place. If the object is of gold or silver this is particularly important since it is the

archaeologists' evidence which helps determine if it is Treasure Trove. Please don't think that archaeologists will demand that you hand over your finds: our first concern is to record discoveries on maps and by drawings so that we have information about archaeological sites all over Norfolk. Please report your discoveries to The Archaeology Dept., Norfolk Museum Service, Castle Museum, Norwich. (Tel: Norwich 22233 Ext. 634 or 635; Weekends or after 5 p.m. messages only on Norwich 21154) or to The Norfolk Archaeological Unit, Union House, Gressenhall, Dereham. (Tel: Gressenhall 528 or 529).... REMEMBER to record where each find was discovered. REMEMBER that pieces of pottery can be as important as metal objects. REMEMBER that digging below the plough-soil will damage an archaeological site. REMEMBER to get the permission of the farmer, or whoever is responsible for the land, before searching. REMEMBER that anything other than gold or silver belongs to the landowner, not to the finder, so make arrangements with him first. REMEMBER that objects of gold or silver are subject to a Treasure Trove Inquest so must be reported. It is illegal to fail to do so. REMEMBER that it is illegal to disturb a site that is a scheduled Ancient Monument."

I must confess that when I read the above words for the first time they left me quite speechless with amazement. I have spent eight frustrating years trying to get archaeologists to adopt this sensible, 20th century attitude towards amateur treasure hunters – alas without success. The reason for this lack of success in the past has been that the *official* view of the Council For British Archaeology (the governing body of Archaeology in Britain) has always been – and still remains as I write these words – that privately owned metal detectors should be banned. Considering that there are now more than 50,000 amateur treasure hunters in Britain it is surely obvious that the CBA is quite out-of-touch with present-day realities. If more county archaeological societies decide to adopt Norfolk's

☐
SOME VICTORIAN SPA TOWNS.
Woods, commons and footpaths around these towns hold vast amounts of pre-1900 losses.

●
SOME VICTORIAN SEASIDE RESORTS with beaches rich in pre-1900 coins and jewellery.

1 Ballater
2 Crieff
3 Stirling
4 Bridge of Earn
5 Ilkley
6 Harrogate
7 Buxton
8 Matlock
9 Leamington
10 Tunbridge Wells
11 Bath
12 Cheltenham
13 Builth Wells
14 Llandrindon Wells
15 Holywell
16 Moffat

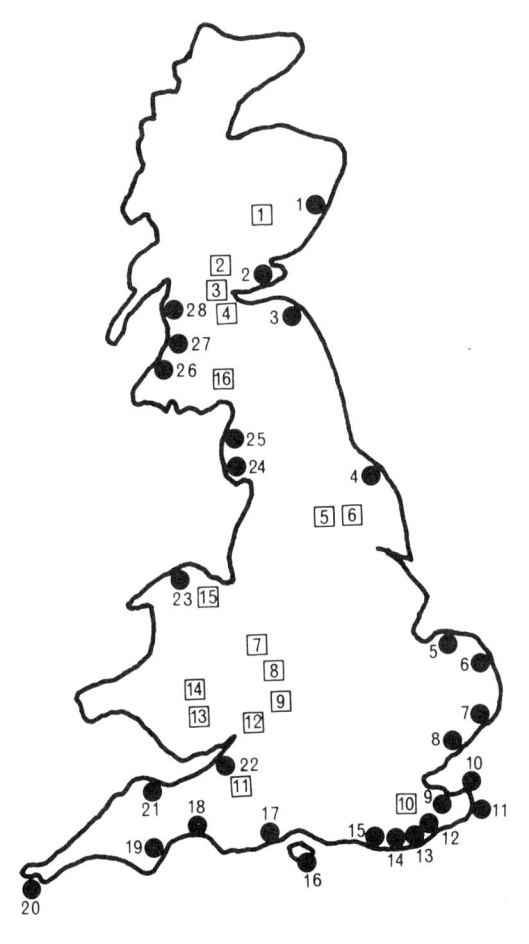

1 Aberdeen
2 Broughty Ferry
3 St. Andrews
4 Scarborough
5 Wells
6 Gt. Yarmouth
7 Clacton
8 Southend
9 Herne Bay
10 Margate
11 Ramsgate
12 Folkestone
13 Hastings
14 Brighton
15 Worthing
16 Ventnor
17 Bournemouth
18 Lyme Regis
19 Torquay
20 Penzance
21 Minehead
22 Weston super Mare
23 Llandudno
24 Southport
25 Blackpool
26 Girvan
27 Ayr
28 Greenock

enlightened approach perhaps the CBA will bring its thinking up-to-date.

CODE OF CONDUCT

It is most important for amateur treasure hunters to follow the rules laid down in the Code of Conduct which are repeated for those who do not know them.

1. Do not interfere with archaeological sites or ancient monuments.
2. Do not leave a mess. By using a sharpened trowel or knife it is simple to extract a coin or other small object from a few inches below the surface without digging a big hole. Replace soil and grass carefully.
3. Keep Britain tidy. Do everyone a favour by placing bottle tops, silver paper, cans etc., in a litter bin. This helps the community and prevents your digging up the same junk again next year.
4. Do not trespass. Always obtain permission to enter private land.
5. Report all unusual historical finds to your local museum and get expert help if you think you have found a site of archaeological interest.
6. Learn the Treasure Trove Laws and report all finds of gold and silver objects to the police.
7. Respect the Country Code. Close all gates when crossing fields and do not frighten animals or damage crops.
8. Be friendly. Always take the opportunity to improve public relations by showing and explaining your detector to anyone who asks.
9. Introduce yourself to other treasure hunters you meet. You could teach each other a lot.
10. Remember you are an ambassador for the whole treasure hunting fraternity. Do not give us a bad name.

TREASURE TROVE

All amateur treasure hunters can benefit from the Treasure

Trove laws and I urge you to seek their protection if you find objects of gold or silver. If an amateur treasure hunter finds a hoard of gold or silver coins, or other objects made from these metals he must immediately report the find to the police or local museum who will take charge of the find and inform the Coroner. At a subsequent inquest the Coroner will decide if the objects were hidden by the owner with intention of recovery at a later date, and who found the objects.

If you found ancient gold or silver coins, for example, it is likely that the person who buried them intended to recover them. In which case the coroner would be most likely to declare the find Treasure Trove. The objects would then become the property of the crown and you, as finder, would receive a reward equal to the full current market value of the find. If not required by a museum the objects would be returned to you in place of a reward or sold on your behalf for the best price obtainable.

If the gold and silver objects were modern the police would endeavour to trace the owner and return the property to him. In this case you would receive no reward unless the owner chose to reward you. If the owner were not traced the Coroner could declare the find Treasure Trove and you would receive the full market value reward.

The Coroner may decide a reward should be shared between a number of treasure hunters who were working together unless they had made an agreement beforehand that finds would not be shared.

If a Coroner decided a find was not Treasure Trove, because the objects did not appear to have been hidden with any intention of recovery, the owner of the property on which the find was made would have a claim to ownership. This would arise in the case of a single coin or other object lost or thrown away. If you had no permission to search the land the Coroner would probably order the find to be given to the owner of the property. It is therefore important to obtain permission before searching property and to sign an agreement to share finds or rewards.

ARCHAEOLOGICAL FINDS

Ancient coins and other objects made from copper, bronze or any other base metal are not Treasure Trove and need not be reported to the Coroner. You should however report them to your local museum and I also recommend you to donate them to the museum.

MODERN FINDS

You are likely to find modern coins, rings, keys, jewellery etc., on your treasure hunting expeditions. It is your duty to give the owners every opportunity to claim the objects you find and you can do this by handing them in at the local police station, advertising in the local press, or pinning up a notice at your house. Common sense should prevail and your actions should take account of the value of the objects.

As a general guide I suggest that all purses, wallets, valuable jewellery and keys be handed to the police from whom the finder may claim them if the owner is not found. Modern coins, inexpensive jewellery and other finds of little value may be retained but you should make sure that local people know you are a treasure hunter and search local sites. If they lose small objects they will know they can contact you with a good chance of recovering the object.

GOLD PROSPECTING

Gold mining rights are vested in the Crown and you must obtain a Crown Estate permit and the permission of the land owner before you pan for gold in a mountain stream.

WRECK MATERIAL

Objects washed ashore from wrecks should be handed to the Coastguard who will return them to you if unclaimed.

Thoughtful Research Pays

Below top: Coins lost on loamy ground with lots of earthworms in it sink rapidly – but are constantly moved by worm action.

Below bottom: Coins lost on stony ground do not sink to great depth – unless there is a build-up of soil on top of them.

Amateur treasure hunters are primarily concerned with the recovery of LOST coins and relics from the top twelve inches of the earth's surface – from loam, sand, silt, and clay to which nature has added amounts of gravel, pebbles, leaf mould, peat, and other materials. Permutations of this "make up" of ground determine how deeply a coin lost two or three hundred years ago will have been buried. On tidal riverbanks and beaches the forces of nature (tides, currents, and wind) disturb the surface material to so great an extent that a three-hundred-year-old coin might lie on the surface one day, be buried beneath three feet of sand or mud the next day, only to return to the surface once more a day later. On arable land a ploughshare can throw a coin up to

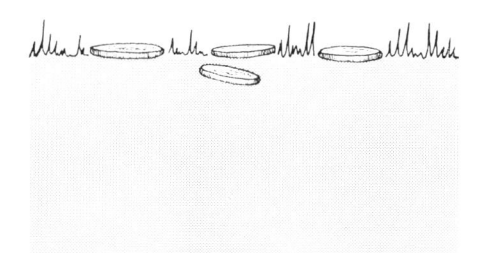

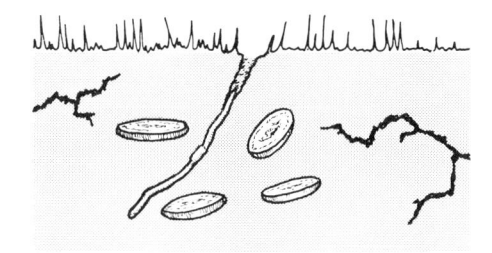

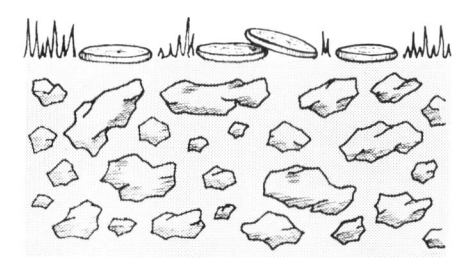

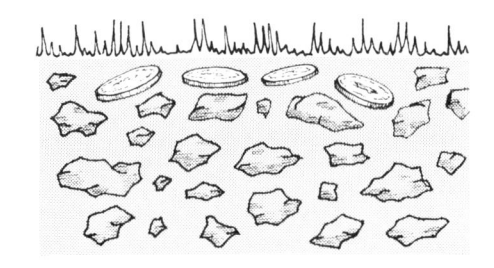

Think constantly about the people who lost the items you are hoping to find. Then, using your knowledge of how coins are moved in the ground, concentrate on the productive areas.

the surface from eighteen inches – and subsequent cultivations (harrowing, rotavations, etc.) can bury it once more in a matter of weeks. Even on grassland undisturbed by man for centuries there is a constant movement up-and-down in the topsoil where earthworm populations, counted at millions per acre, continually aerate the soil and churn it far more efficiently than man's machines. Add to all this the forces of erosion, weathering, and soil-creep and you will begin to realise that our "territory" is never still, and that the "treasures" we hope to find are forever on the move. (I hope you will also begin to realise how ridiculous is the claim by archaeologists that our activities damage stratification.)

In order to succeed at amateur treasure hunting, especially when you rate success as finding the oldest coins, you must take these natural movements into account when selecting your sites and working with your detector. If you plan to work a beach WAIT until nature provides the best conditions – during and immediately after heavy seas which turn over the sand to a depth of three feet. If you plan to work a tidal riverside WAIT – for low tides following heavy rains when the riverbed pattern will be completely changed and YOU will be the first treasure hunter to work the "new" foreshore. If research indicates that a certain farm field constantly throws up Roman coins WAIT – until the farmer has ploughed it deeply in the autumn and brought material beyond the range of your detector to the surface. If an old meadow used as a playing field in Victorian times is your chosen site WAIT – until rainy weather stirs the earthworms into action and some of the coins lying at ten and twelve inches are moved an inch or two towards the surface and WITHIN RANGE of your detector.

It is knowledge such as this – and the patience to work *with* nature – that brings success.

Every time you survey a new field or patch of grass you should examine the topsoil carefully. Are there lots of pebbles and stones in the "make up"? If YES, then coins and relics are likely to be near the surface even if they have been in the ground three hundred years because pebbles and stones hold up the coins, preventing them from sinking deeply. Is it an "acid" soil – the type you might find in highland areas? If YES, then it will hold few earthworms because they cannot live in acid soils. If there has also been a build-up of surface material – peat, leafmould, grass-cuttings – then any coins and relics in the ground will be deeply buried – perhaps beyond the range of your equipment.

Let me say once again that your own notebooks and record-of-finds, compiled during a year or so of amateur treasure hunting, will be worth their weight in gold. By comparing different types of ground, sites, and situations you will be able to eliminate hours of fruitless searching and go DIRECTLY to the spots where coins are within reach. You should also make notes, draw sketches and diagrams, and RECORD your finds when working stream banks, hill slopes, valleys, in fact wherever you go with your detector. Experience will teach you to look at a scene and to see it as it was three hundred years ago. Ask yourself how people crossed a particular ford, negotiated a steep slope, found their way across a boggy river valley in those days. And how has the lie of land and water changed with the passing centuries? Where – in those days – were the places PEOPLE lost their posessions? WHEN QUESTIONS LIKE THIS SPRING TO MIND AUTOMATICALLY YOU ARE ON YOUR WAY TO TREASURE HUNTING SUCCESS.

Productive sites merit careful lines-and-pins searches. Only a small area of your detector's search head penetrates the ground to maximum depth. You can miss valuable coins even during a slow and methodical search. If a site yields high value coins and relics one day, go back several times and rework the site.

Useful Information

"Treasure Hunting",
N.P.C.
Sovereign House,
Brentwood,
Essex.
Monthly magazine which covers all aspects of treasure hunting.

"Antique Bottle Collecting",
Chapel House Farm,
Newport Road,
Albrighton,
Nr. Wolverhampton.
Monthly magazine on the hobbies of collecting bottles and Victorian relics.

Old Bottle Club of Great Britain,
14 Derwent Crescent,
Whitehill, Kidsgrove,
Staffordshire.
Monthly meetings, quarterly magazine, sponsored shows.

Weekend Treasure Hunters,
104, Harwal Road,
Redcar,
Cleveland.
Edward Fletcher's club for treasure hunting enthusiasts.

Coin Monthly,
N.P.C.
Sovereign House,
Brentwood, Essex.
Britains best selling coin magazine which includes treasure hunting news.

Gemcraft,
P.O. Box 35,
Hemel Hempstead, Herts.
Monthly magazine covering rockhounding, lapidary, treasure hunting.

METAL DETECTOR MANUFACTURERS

C-Scope Metal Detectors,
Wooton Road,
Ashford,
Kent.

Joan Allen,
184 Main Road,
Biggin Hill,
Kent.

Savo Electronics Ltd.,
Longman Road,
Longman Industrial Estate,
Inverness.

Altek Instruments,
1 Green Lane,
Walton on Thames,
Surrey.

Araldo Electronics,
P.O. Box 10,
Banstead,
Surrey.

Young Electronics Ltd.,
184 Royal College Street,
London NW1 9NN

Best Instrument Design Ltd.,
13–16 London Road,
Tunbridge Wells,
Kent.

A.D. Detectors,
Garage Lane,
Setchley,
Kings Lynn.

Sol Invictus Detectors Ltd.,
81 Moorgate Street,
Blackburn,
Lancs.

Pulse Induction Ltd.,
42 Notting Hill Gate,
London W11 3HX

MAIN RETAILERS OF TREASURE HUNTING EQUIPMENT

Treasure Hunting Supplies,
30 Roundwood Road,
Willesdon
London NW 10.

Treasure Hunters Centre,
221 Lakey Lane,
Hall Green,
Birmingham.

Treasure Hunters Centre,
32 Craven Street,
London WC2.

Pieces of Eight,
155 Robert Street,
London NW1.

Seaway,
10 Beach Road,
Perranporth,
Cornwall.

Greens Metal Detectors,
110 Crwys Road,
Cardiff.

Home Counties Treasure
  Hunters Shop,
344 St. Albans Road,
Watford.

Tape Centre,
83 Alfreton Road,
Nottingham.

Evingtons Hobby Crafts,
5 Alexandra Road,
Grimsby.

Treasure Hunters Centre,
83/85 Lark Lane,
Liverpool.

Lapidary Supplies,
330 Gloucester Road,
Horfield,
Bristol.

Renwicks,
76/78 Upper High Street,
Ryde,
Isle of Wight.

Trade Winds Craft Studios,
4 High Street,
Caister-on-sea,
Great Yarmouth.

Gemset,
31 Albion Street,
Broadstairs,
Kent.

Les Ward,
53 Sheffield Road,
Chesterfield.

Tuckton Garden Centre,
133 Tuckton Road,
Tuckton,
Bournemouth.

BOTTLE COLLECTING PRICE GUIDE

By Gordon Litherland. Over 1500 bottles priced. Over 100 illustrations and photographs. Chapters on finding, identifying, cleaning, dating bottles, plus methods of manufacture and a dictionary of bottle collecting terms. £1.95.

COLLECTING POT-LIDS

Coloured, black and white with current price trends by A. Ball. Abe Ball is the country's leading authority on Pratt-ware, and this book covers collecting both coloured and black and white pot-lids. As well as the author's advice on collecting and many illustrations, there is a price guide covering almost 1,000 different lids. Suitable for established collectors and beginners as well as dealers. Full of illustrations and photographs. £1.95.

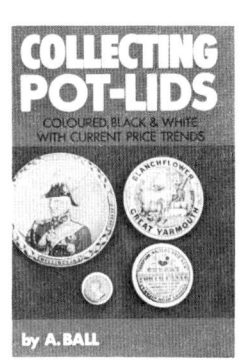

COLLECTING POTTERY

Underglaze printed ware – with price guide by A. Ball.
The products of all the main factories which produced underglaze colour-printed ware are listed, illustrated and priced. The range extends from the plates, mugs, jugs, cups, saucers, pot-lids of the nineteenth century to the more common mass produced ginger beer bottles and cream pots. An invaluable price guide for collectors. £1.95.

SEALED BOTTLES

Their History and Evolution (1630–1930) by Roy Morgan. 150 black and white illustrations (including 50 full plate photographs). This is the first book devoted solely to sealed bottles of this period since "Sealed Bottles" by Sheelagh Ruggles-Brise (Country Life 1949).
Sections on history, development, means of dating. Descriptions of each type and details of specific bottles and their history. £4.50.

Available from your booksellers or the publishers (post free)

M.A.B. PUBLISHING
801 Burton Road, Midway, Burton upon Trent, Staffs, DE11 0DN.
Tel. 0283–66255

Acknowledgements

I wish to thank the following for their assistance in supplying information and material for this book:

Photographs: Ian Bennett, A.I.I.P., Burton upon Trent.
Gordon Litherland, Burton upon Trent.
Derek Lovatt, Burton upon Trent.
Archie Eduljee, Burton upon Trent.
Joan Allen Electronics Ltd (Pages 56, 59, 61).
Young Electronics Ltd. (Page 21).
Roger Green, "Antique Bottle Collecting".

Drawings: Alison and Charles Griffin, Maidstone.

Archaeology Department, Norfolk Museum Service.